Numerical Simulation of Optical Wave Propagation

With examples in MATLAB®

Library of Congress Cataloging-in-Publication Data

Schmidt, Jason Daniel, 1975-
Numerical simulation of optical wave propagation with examples in MATLAB / Jason D. Schmidt.
p. cm. -- (Press monograph ; 199)
Includes bibliographical references and index.
ISBN 978-0-8194-8326-3
1. Optics--Mathematics. 2. Wave-motion, Theory of--Mathematical models. 3. MATLAB. I. Title.
QC383.S36 2010
535'.42015118--dc22
2010015089

Published by

SPIE
P.O. Box 10
Bellingham, Washington 98227-0010 USA
Phone: +1 360.676.3290
Fax: +1 360.647.1445
Email: Books@spie.org
Web: http://spie.org

Printed in the United States of America.

Fourth printing: 2017

About the cover: 50-watt laser for generating mesospheric sodium guide stars over 90 km above the ground. In operation at the Air Force Research Laboratory's 3.5-m telescope at the Starfire Optical Range, Kirtland AFB, NM. (Robert Q. Fugate, © 2005, Albuquerque, NM).

Numerical Simulation of Optical Wave Propagation

With examples in MATLAB®

Jason D. Schmidt

SPIE PRESS
Bellingham, Washington USA

Contents

Preface

Diffraction is a very interesting and active area of optical research. Unfortunately, analytic solutions are rare in many practical problems, particularly when optical waves propagate through randomly fluctuating media. For many of these problems, researchers must resort to numerical solutions. Still, simulations in optical diffraction are challenging. Usually, these simulations take advantage of discrete Fourier transforms, which means using discretely spaced samples on a finite-sized grid. This leads to a few tradeoffs in speed and memory versus accuracy. Thus, the parameters of the sampling grids must be chosen very carefully. Some people seek to fully automate those choices, but this cannot be done automatically in every case. To determine grid properties, one must carefully consider computational speed, available computer memory, the Nyquist sampling criterion, geometry, accurate representation of source apertures, and impact on the propagated field's quantities of interest.

This book grew out of an independent study I did while I was a doctoral student at University of Dayton. The study was directed by LtCol Matthew Goda, then a professor at the Air Force Institute of Technology (AFIT). After the independent study was over, Goda then created a course at AFIT on wave-optics simulations. When I graduated, I became a professor at AFIT while Goda moved on to a new military assignment. When I began teaching the wave-optics simulation course, there was no book written to the level of detail required for a graduate course focused on wave-optics simulations and sampling requirements. The course was always taught out of the professor's notes, originally compiled by Goda. Compiling these notes was no small feat, and Goda did a tremendous job combining material from books on discrete Fourier transforms, optics journal articles and conference proceedings, technical reports from companies like the Optical Sciences Company and MZA Associates Corporation, and private communication with researchers.

Until this book, simulations have always been an afterthought in just a few books on image processing and nonlinear optics. Clearly there was a gap between the practical knowledge required to perform wave-optics simulations and the theoretical material covered in great Fourier-optics textbooks like those by Joseph Goodman and Jack Gaskill. I have heard professors across the U.S. talk about how they include material on simulations in their graduate Fourier-optics courses. I applaud them for that effort because it is challenging to teach students both the theory and practical simulation of Fourier optics in one course. However, if the students are to become capable enough to write wave-optics simulations for thesis or dissertation research and beyond, they cannot get enough detail in a one-term Fourier-optics course. This is why AFIT has separate courses on Fourier optics and wave-optics simulations.

This book is intended for graduate students in programs like physics, electrical engineering, electro-optics, or optical science. The book gives all of the relevant equations from Fourier optics, but to fully understand and appreciate the material, it is important to have a thorough understanding of Fourier optics before reading this book.

I believe that part of the benefit of this book is the use of specific code examples, rather than just pseudo-code. However, the programming or scripting language for the examples needs to be one that is widely used and easy to understand by those who do not already use it. For those reasons, I have used MATLAB in all of the examples throughout this book. It is heavily used in engineering both at universities and research institutions. Further, it is easy to read because of its simple language and because many numerical algorithms, such as discrete Fourier transforms and convolution, are part of its basic library. If I used other languages like C, C++, FORTRAN, Java, and Python, I would need to pick a particular external library of numerical routines or write my own algorithms and include them in the book. I believe that using MATLAB in this book allows readers to focus on the wave propagation, rather than the most basic numerical algorithms like discrete Fourier transforms. Further, any user with access to the MATLAB interpreter can execute the code examples as shown. No additional libraries need to be acquired and installed. Moreover, my examples rarely use MATLAB's toolboxes, relying heavily on its basic functionality. Readers should note that the code examples used throughout the book are designed for conceptual simplicity, rather than optimized for speed or memory usage. I encourage readers to rework my MATLAB examples to achieve greater performance or even implement them in other languages.

I offer my thanks and appreciation to all those who have paved the way for this work, particularly Glenn Tyler, David Fried, and Phillip Roberts at the Optical Sciences Company and Steve Coy at MZA Associates Corporation. In 1982, Fried and Tyler wrote a technical report describing methods of simulating optical wave propagation and related sampling constraints. A few years later, Roberts wrote a follow-on report giving another clear, nicely detailed description of one-step, two-step, and angular spectrum propagation methods. More recently, Coy wrote a technical report that gives a very nice description of the relationship between sampling requirements propagation geometry. These reports formed the beginnings of Goda's notes and eventually this book.

Also, thanks to those who answered my questions about wave-optics simulations while I was a student at UD and then while I taught the wave-optics simulation course as a professor at AFIT: Jeffrey Barchers, Troy Rhoadarmer, Terry Brennan, and Don Link. These gentlemen are experienced and accomplished researchers whose advice was very much appreciated. Additionally, thanks to Michael Havrilla for his help with the basic electrodynamics in Ch. 1.

Special thanks to Matthew Goda for his foundational work in the course and its notes. Without him, this book would not be possible. He made much of the material in this book accessible to dozens of students who went on to do great things for the U.S. Air Force. Finally, I'd like to thank all those students who helped find errors in the drafts of this book and whose inquisitive nature caused me to refine and add material along the way.

Jason Schmidt
June, 2010

The MATLAB® code for the tools and examples used in the book is available for download here:
http:/spie.org/Samples/Pressbook_Supplemental/PM199_sup.zip.

Chapter 1
Foundations of Scalar Diffraction Theory

Light can be described by two very different approaches: classical electrodynamics and quantum electrodynamics. In the classical treatment, electric and magnetic fields are continuous functions of space and time, and light comprises co-oscillating electric and magnetic wave fields. In the quantum treatment, photons are elementary particles with no mass nor charge, and light comprises one or more photons. There is rigorous theory behind each approach, and there is experimental evidence supporting both. Neither approach can be dismissed, which leads to the wave-particle duality of light. Generally, classical methods are used for macroscopic properties of light, while quantum methods are used for submicroscopic properties of light.

This book describes macroscopic properties, so it deals entirely with classical electrodynamics. When the wavelength λ of an electromagnetic wave is very small, approaching zero, the waves travel in straight lines with no bending around the edges of objects. That is realm of geometric optics. However, this book treats many situations in which geometric optics are inadequate to describe observed phenomena like diffraction. Therefore, the starting point is classical electrodynamics with solutions provided by scalar diffraction theory. Geometric optics is treated briefly in Sec. 6.5.

1.1 Basics of Classical Electrodynamics

Classical electrodynamics deals with relationships between electric fields, magnetic fields, static charge, and moving charge (i.e., current) in space and time based on the macroscopic properties of the materials in which the fields exist. We define each quantity here along with some basic relationships. This introduces the reader to the quantities in Maxwell's equations, which describe how electrically charged particles and objects give rise to electric and magnetic fields. Maxwell's equations are introduced here in their most general form, and then the discussion focuses on a specific case and solutions for oscillating electric and magnetic fields, which light comprises.

1.1.1 Sources of electric and magnetic fields

Electric charge, measured in coulombs, is a fundamental property of elementary particles and bulk materials. Classically, charge may be positive, negative, or zero. Further, charge is quantized, specifically the smallest possible nonzero amount of charge is the *elementary charge* $e = 1.602 \times 10^{-19}$ C. All nonzero amounts of charge are integer multiples of e. For bulk materials, the integer may be very large so that total charge can be treated as continuous rather than discrete. We denote the volume density of free charge, measured in coulombs per cubic meter, by $\rho(\mathbf{r}, t)$, where $\mathbf{r}$ is a three-dimensional spatial vector, and t is time. Moving charge density is called free volume current density $\mathbf{J}(\mathbf{r}, t)$. Volume current density is measured in Ampères per square meter (1 A = 1 C/s). This represents the time rate at which charge passes through a surface of unit area. Finally, charge is conserved, meaning that the total charge of any system is constant. This is mathematically stated by the continuity equation

$$\nabla \cdot \mathbf{J}(\mathbf{r}, t) + \frac{\partial \rho(\mathbf{r}, t)}{\partial t} = 0. \tag{1.1}$$

Almost every material we encounter in life is composed of many, many atoms each with many positive and negative charges. Usually, the numbers of positive and negative charges are equal or nearly equal so that the whole material is electrically neutral. Still, such a material can give rise to electric or magnetic fields when the total charge and free current are zero. If the distribution of charge is not homogeneous or if the charges are circulating in tiny current loops, fields could be present.

The separation of charge is described by the electric dipole moment, which is the amount of separated charge times the separation distance. If a bulk material has its charge arranged in many tiny dipoles, it is said to be electrically polarized. The volume polarization density $\mathbf{P}(\mathbf{r}, t)$ is the density of electric dipole moments per unit volume, measured in coulombs per square meter.

Magnetization is a similar concept for moving charge. Charge circulating in a tiny current loop is described by magnetic dipole moment, which is the circulating current times the area of the loop. When a bulk material has internal current arranged in many tiny loops, it is said to be magnetized. The volume magnetization density $\mathbf{M}(\mathbf{r}, t)$ is the density of magnetic dipole moments per unit volume, measured in Ampères per meter.

1.1.2 Electric and magnetic fields

When a hypothetical charge, called a test charge, passes near a bulk material that has non-zero ρ, $\mathbf{J}$, $\mathbf{P}$, or $\mathbf{M}$, the charge experiences a force. This interaction is characterized by two vectors $\mathbf{E}$ and $\mathbf{B}$. The electromagnetic force $\mathbf{F}$ on a test particle at a given point and time is a function of these vector fields and the particle's charge q and velocity $\mathbf{v}$. The Lorentz force law describes this interaction as

$$\mathbf{F} = q(\mathbf{E} + \mathbf{v} \times \mathbf{B}). \tag{1.2}$$

If this empirical statement is valid (and, of course, countless experiments over the course of centuries have shown that it is), then two vector fields $\mathbf{E}$ and $\mathbf{B}$ are thereby defined throughout space and time, and these are called the "electric field" and "magnetic induction."[1]

Eq. (1.2) can be examined in a little more detail to provide more intuitive definitions of these fields. The electric field is the amount of force per unit of test charge when the test charge is stationary, given by

$$\mathbf{E} = \lim_{q \to 0^+} \left. \frac{\mathbf{F}}{q} \right|_{\mathbf{v}=\mathbf{0}} . \tag{1.3}$$

This is called a push-and-pull force because the force is in either the same or opposite direction as the field, depending on the sign of the charge. Electric field is measured in units of volts per meter (1 V = 1 N m/C). The magnetic field is related to the amount of force per unit test charge given by

$$\mathbf{v} \times \mathbf{B} = \lim_{q \to 0^+} \left. \frac{\mathbf{F} - q\mathbf{E}}{q} \right|_{\mathbf{v} \neq \mathbf{0}} . \tag{1.4}$$

The force due to a magnetic field is called deflective because it is perpendicular to the particle's velocity, which deflects its trajectory. Magnetic field is measured in units of Tesla [1 T = 1 N s/(C m)].

With this understanding of the fields, they now need to be related to the sources. This was accomplished through centuries of experimental measurements and theoretical and intuitive insight, resulting in

$$\nabla \times \mathbf{E} + \frac{\partial \mathbf{B}}{\partial t} = \mathbf{0} \tag{1.5}$$

$$\nabla \times \mathbf{B} - \mu_0 \epsilon_0 \frac{\partial \mathbf{E}}{\partial t} = \mu_0 \left(\mathbf{J} + \frac{\partial \mathbf{P}}{\partial t} + \nabla \times \mathbf{M} \right) . \tag{1.6}$$

These are two of Maxwell's equations, the former being Faraday's law and the latter being Ampère's law with Maxwell's correction. In Eq. (1.6), the sources on the right hand side include the free current $\mathbf{J}$ and two terms due to bound currents. These are the polarization current $\partial \mathbf{P}/\partial t$ and the magnetization current $\nabla \times \mathbf{M}$.

These equations can be written in a more functionally useful form. Eq. (1.6) can be rewritten as

$$\nabla \times \left(\frac{\mathbf{B}}{\mu_0} - \mathbf{M} \right) = \mathbf{J} + \frac{\partial}{\partial t} \left(\epsilon_0 \mathbf{E} + \mathbf{P} \right) . \tag{1.7}$$

Making the definitions

$$\mathbf{D} = \epsilon_0 \mathbf{E} + \mathbf{P} \tag{1.8}$$

$$\mathbf{H} = \frac{\mathbf{B}}{\mu_0} - \mathbf{M} \tag{1.9}$$

introduces the concepts of electric displacement $\mathbf{D}$ and magnetic field $\mathbf{H}$, which are fields that account for the medium's response to the applied fields. Now, the working form of these Maxwell equations becomes

$$\nabla \times \mathbf{E} = -\frac{\partial \mathbf{B}}{\partial t} \tag{1.10}$$

$$\nabla \times \mathbf{H} = \mathbf{J} + \frac{\partial \mathbf{D}}{\partial t}. \tag{1.11}$$

Further, when these are combined with conservation of charge expressed in Eq. (1.1), this leads to

$$\nabla \cdot \nabla \times \mathbf{H} = \nabla \cdot \mathbf{J} + \frac{\partial}{\partial t}\nabla \cdot \mathbf{D} \tag{1.12}$$

$$= -\frac{\partial \rho}{\partial t} + \frac{\partial}{\partial t}\nabla \cdot \mathbf{D} \tag{1.13}$$

$$= 0. \tag{1.14}$$

Focusing on the right-hand side,

$$\frac{\partial}{\partial t}(\nabla \cdot \mathbf{D} - \rho) = 0 \tag{1.15}$$

$$\nabla \cdot \mathbf{D} - \rho = f(\mathbf{r}), \tag{1.16}$$

where $f(\mathbf{r})$ is an unspecified function of space but not time. Causality requires that $f(\mathbf{r}) = 0$ before the source is turned on, yielding Coulomb's law:

$$\nabla \cdot \mathbf{D} = \rho. \tag{1.17}$$

Similar manipulations yield

$$\nabla \cdot \mathbf{B} = 0. \tag{1.18}$$

This indicates that magnetic monopole charges do not exist. Finally, Eqs. (1.10), (1.11), (1.17), and (1.18) constitute Maxwell's equations.[1]

In this model of macroscopic electrodynamics, Eqs. (1.10) and (1.11) are two independent vector equations. With three scalar components each, these are six independent scalar equations. Unfortunately, given knowledge of the sources, there are four unknown vector fields $\mathbf{D}$, $\mathbf{B}$, $\mathbf{H}$, and $\mathbf{E}$. Each has three scalar components for a total of twelve unknown scalars. With so many more unknown field components than equations, this is a poorly posed problem.

The key is to understand the medium in which the fields exist. This produces a means of relating $\mathbf{P}$ to $\mathbf{E}$ and $\mathbf{M}$ to $\mathbf{H}$, which amount to six more scalar equations. For example, in simple media (linear, homogeneous, and isotropic),

$$\mathbf{P} = \epsilon_0 \chi_e \mathbf{E} \tag{1.19}$$

$$\mathbf{M} = \chi_m \mathbf{H}, \tag{1.20}$$

where χ_e is the electric susceptibility of the medium and χ_m is its magnetic susceptibility. Substituting these into Eqs. (1.8) and (1.9) yields

$$\begin{aligned} \mathbf{D} &= \epsilon_0 \mathbf{E} + \mathbf{P} &(1.21)\\ &= \epsilon_0 (1 + \chi_e) \mathbf{E} &(1.22)\\ &= \epsilon \mathbf{E} &(1.23) \end{aligned}$$

and

$$\begin{aligned} \mathbf{B} &= \mu_0 (\mathbf{H} + \mathbf{M}) &(1.24)\\ &= \mu_0 (1 + \chi_m) \mathbf{H} &(1.25)\\ &= \mu \mathbf{H}, &(1.26) \end{aligned}$$

where $\epsilon = (1 + \chi_e)\,\epsilon_0$ is the electric permittivity and $\mu = (1 + \chi_m)\,\mu_0$ is the magnetic permeability of the medium. Now, this simplifies Eqs. (1.10) and (1.11) so that

$$\nabla \times \mathbf{E} = -\mu \frac{\partial \mathbf{H}}{\partial t} \quad (1.27)$$

$$\nabla \times \mathbf{H} = \mathbf{J} + \epsilon \frac{\partial \mathbf{E}}{\partial t}. \quad (1.28)$$

Now, there are still six equations but only six unknowns (as long as the free current density $\mathbf{J}$ is known). Finally, with a proper understanding of the materials, this is a well posed problem.

1.2 Simple Traveling-Wave Solutions to Maxwell's Equations

There are many solutions to Maxwell's equations, but there are only a few that can be written in closed form without an integral. This section begins with transforming Maxwell's four equations into two uncoupled wave equations. It continues with a few specific simple solutions such as the infinite-extent plane wave. A more general solution is left to the next section.

1.2.1 Obtaining a wave equation

This book deals with optical wave propagation through linear, isotropic, homogeneous, nondispersive, dielectric media in the absence of source charges and currents. In this case, the media discussed throughout the remainder of this book have

$$\begin{aligned} &\epsilon = \text{a scalar, independent of } \lambda, \mathbf{r}, t &(1.29)\\ &\mu = \mu_0 &(1.30)\\ &\rho = 0 &(1.31)\\ &\mathbf{J} = \mathbf{0}. &(1.32) \end{aligned}$$

Taking the curl of Eq. (1.27) yields

$$\nabla \times (\nabla \times \mathbf{E}) = -\mu_0 \frac{\partial}{\partial t} (\nabla \times \mathbf{H}) . \tag{1.33}$$

Then, substituting in Eq. (1.28) gives

$$\nabla \times (\nabla \times \mathbf{E}) = -\mu_0 \epsilon \frac{\partial^2}{\partial t^2} \mathbf{E}. \tag{1.34}$$

Now, applying the vector identity $\nabla \times (\nabla \times \mathbf{E}) = \nabla (\nabla \cdot \mathbf{E}) - \nabla^2 \mathbf{E}$ leads to

$$\nabla (\nabla \cdot \mathbf{E}) - \nabla^2 \mathbf{E} = -\mu_0 \epsilon \frac{\partial^2}{\partial t^2} \mathbf{E}. \tag{1.35}$$

Finally, substituting in Eqs. (1.17) and (1.23), and keeping in mind that ϵ is independent of position results in a wave differential equation:

$$\nabla^2 \mathbf{E} - \mu_0 \epsilon \frac{\partial^2}{\partial t^2} \mathbf{E} = \mathbf{0}. \tag{1.36}$$

Similar manipulations beginning with the curl of Eq. (1.28) yield

$$\nabla^2 \mathbf{B} - \mu_0 \epsilon \frac{\partial^2}{\partial t^2} \mathbf{B} = \mathbf{0}. \tag{1.37}$$

When the Laplacian is used on the Cartesian components of $\mathbf{E}$ and $\mathbf{B}$, the result is six uncoupled but identical equations of the form

$$\left(\nabla^2 - \mu_0 \epsilon \frac{\partial^2}{\partial t^2} \right) U(x, y, z) = 0, \tag{1.38}$$

where the scalar $U(x, y, z)$ stands for any of the x-, y- or z- directed components of the vector fields $\mathbf{E}$ and $\mathbf{B}$.

At this point, we can define index of refraction

$$n = \sqrt{\frac{\epsilon}{\epsilon_0}} \tag{1.39}$$

and the vacuum speed of light

$$c = \frac{1}{\sqrt{\mu_0 \epsilon_0}} \tag{1.40}$$

so that

$$\left(\nabla^2 - \frac{n^2}{c^2} \frac{\partial^2}{\partial t^2} \right) U(x, y, z) = 0. \tag{1.41}$$

The electric and magnetic fields that compose light are traveling wave fields. Therefore, fields with harmonic time dependence $\exp(-i2\pi\nu t)$ (where ν is the wave frequency) are the types of solutions sought for the purposes of this book. When this is substituted into Eq. (1.41), the result is

$$\left[\nabla^2 + \left(\frac{2\pi n\nu}{c}\right)^2\right] U = 0. \tag{1.42}$$

Typically, the wavelength is given by $\lambda = c/\nu$, and the wavenumber is defined as $k = 2\pi/\lambda$ so that

$$\left[\nabla^2 + k^2 n^2\right] U = 0. \tag{1.43}$$

This is the Helmholtz equation, and it appears in many other branches of physics including thermodynamics and quantum mechanics. At this point, we can dispense with the time dependence since it is the same for all solutions of the Helmholtz equation. From this point forward, the field $U\left(x, y, z\right)$ refers to the phasor portion of the optical field (i.e, no time dependence). Further, we define the units of $U\left(x, y, z\right)$ to be square-root watts per meter (1 W = 1 J/s = 1 N m/s) so that optical irradiance $I = |U|^2$ is in units of watts per meter squared. The value of the electric field or magnetic induction can always be obtained by a simple conversion of units.

1.2.2 Simple traveling-wave fields

There are several simple traveling-wave fields that are useful in this book. These are planar, spherical, and Gaussian-beam waves. With each of these solutions, the field at all points always maintains its planar, spherical, or Gaussian-beam form, and parameters like radius of curvature change in a simple manner as the wave propagates. The next section on scalar diffraction theory handles more general cases.

A planar wave is the simplest possible traveling wave. It has uniform amplitude and phase in any plane perpendicular to its direction of propagation. More generally, when the optical axis is not along the direction of propagation, a planar wave field is given by

$$U_P\left(\mathbf{r}\right) = A \exp\left(i\mathbf{k} \cdot \mathbf{r}\right), \tag{1.44}$$

where A is the amplitude of the wave and

$$\mathbf{k} = \frac{2\pi}{\lambda}\left(\alpha\hat{\mathbf{x}} + \beta\hat{\mathbf{y}} + \gamma\hat{\mathbf{z}}\right) \tag{1.45}$$

is the wavevector with direction cosines given by α, β, and γ. Then, making the direction cosines more explicit,

$$U_P\left(\mathbf{r}\right) = A \exp\left[i\frac{2\pi}{\lambda}\left(\alpha x + \beta y + \gamma z\right)\right]. \tag{1.46}$$

This wave travels at an angle $\cos^{-1}\alpha$ from the x-axis and $\cos^{-1}\beta$ from the y-axis as shown in Fig. 1.1.

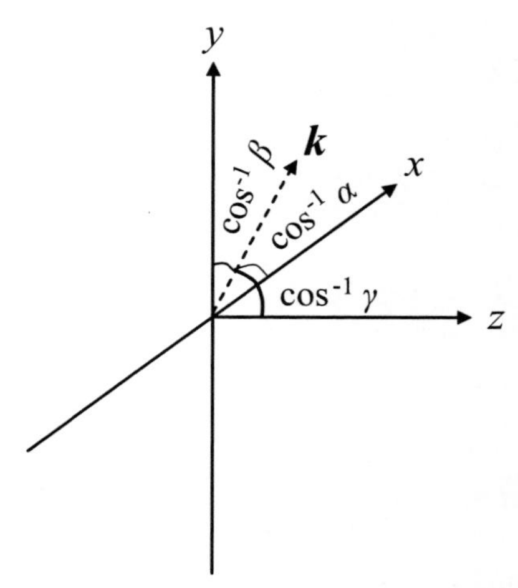

Figure 1.1 Depiction of direction cosines α, β, and γ.

A spherical wave is the next simplest wave field. It has a wavefront that is spherical in shape, and it is either diverging or converging. The energy of the wave is spread uniformly over a spherical surface with area given by $4\pi R^2$, where R is the wavefront radius of curvature. Conservation of energy requires that the amplitude is accordingly proportional to R^{-1}. A spherical wave is given by

$$U_S(\mathbf{r}) = A\frac{\exp[ikR(\mathbf{r})]}{R(\mathbf{r})}. \tag{1.47}$$

If the center of the sphere is located at $\mathbf{r}_c = (x_c, y_c, z_c)$, then at an observation point $\mathbf{r} = (x, y, z)$, the radius of curvature is given by

$$R(\mathbf{r}) = \sqrt{(x - x_c)^2 + (y - y_c)^2 + (z - z_c)^2}. \tag{1.48}$$

Often in optics, attention is restricted to regions of space that are very close to the optical axis. This is called the paraxial approximation, and assuming propagation in the positive z direction, this approximation is mathematically written as

$$\cos^{-1}\alpha \ll 1 \tag{1.49}$$

$$\cos^{-1}\beta \ll 1. \tag{1.50}$$

With this approximation, we eliminate the square root by expanding it as a Taylor series and keeping only the first two terms, yielding

$$R(\mathbf{r}) \simeq \Delta z\left[1 + \frac{1}{2}\left(\frac{x - x_c}{\Delta z}\right)^2 + \frac{1}{2}\left(\frac{y - y_c}{\Delta z}\right)^2\right], \tag{1.51}$$

where we have defined $\Delta z = |z - z_c|$. With the paraxial approximation, a spherical wave is approximately

$$U_S(\mathbf{r}) \simeq A\frac{e^{ik\Delta z}}{\Delta z}e^{i\frac{k}{2\Delta z}\left[(x-x_c)^2+(y-y_c)^2\right]}. \tag{1.52}$$

One final simple traveling wave often encountered in optics is the Gaussian-beam wave. It has a Gaussian amplitude profile and "paraxially spherical" wavefront. The full derivation of the Gaussian-beam solution invokes the paraxial approximation along the way. Such a derivation can be found in common laser textbooks like Refs. 2–3. This solution is given by

$$U_G(\mathbf{r}) = \frac{A}{q(z)} \exp\left[ik\frac{x^2+y^2}{2q(z)}\right], \tag{1.53}$$

where

$$\frac{1}{q(z)} = \frac{1}{R(z)} + \frac{i\lambda}{\pi W^2(z)} \tag{1.54}$$

and the beam radius and wavefront radius of curvature are given by

$$W^2(z) = W_0^2\left[1+\left(\frac{\lambda z}{\pi W_0^2}\right)^2\right] \tag{1.55}$$

$$R(z) = z\left[1+\left(\frac{\pi W_0^2}{\lambda z}\right)^2\right], \tag{1.56}$$

where W_0 is the minimum spot radius. At any point along the z axis, $W(z)$ is the $1/e$ radius of the field amplitude. Also, by this convention, $W(0) = W_0$ so that the minimum spot radius is located at $z = 0$.

1.3 Scalar Diffraction Theory

Often, the optical source is not a simple planar, spherical, nor Gaussian-beam wave. For more general cases, we must use more sophisticated means to solve the scalar Helmholtz equation. This means taking advantage of Green's theorem with clever use of boundary conditions. This process is not discussed in detail here, but the interested reader should consult books like Refs. 4–5 for a detailed treatment.

The geometry for this more general case is shown in Fig. 1.2. In this figure, the coordinates are $\mathbf{r}_1 = (x_1, y_1)$ in the source plane and $\mathbf{r}_2 = (x_2, y_2)$ in the observation plane. The distance between the two planes is Δz. The figure illustrates the basic problem: given the source-plane optical field $U(x_1, y_1)$, what is the observation-plane field $U(x_2, y_2)$? The solution is given by the Fresnel diffraction integral

$$U(x_2, y_2) = \frac{e^{ik\Delta z}}{i\lambda\Delta z} \int_{-\infty}^{\infty}\int_{-\infty}^{\infty} U(x_1, y_1)\, e^{i\frac{k}{2\Delta z}\left[(x_1-x_2)^2+(y_1-y_2)^2\right]}\, dx_1 dy_1. \tag{1.57}$$

Note that this is not the most general solution. In fact, it is a paraxial approximation, but it is general enough and accurate enough for the purposes of this book.

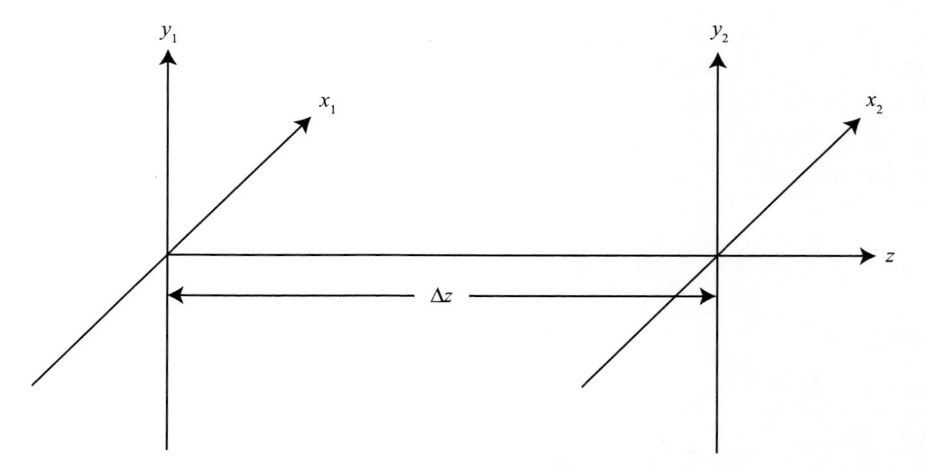

Figure 1.2 Coordinate systems for optical-wave propagation.

There are only a handful of analytic solutions to Eq. (1.57). Particularly, Fresnel diffraction from a rectangular aperture is used many times as an example in Chs. 6–8. Because few other Fresnel diffraction problems have an analytic answer, this one is used to compare against numerical results in several example simulations. When the source field is

$$U(x,y) = \mathrm{rect}\left(\frac{x_1}{D}\right)\mathrm{rect}\left(\frac{y_1}{D}\right), \tag{1.58}$$

(for the definition of the rect function, see Appendix A) the diffracted field in the observation plane a distance Δz away is given by

$$U(x_2,y_2) = \frac{e^{ik\Delta z}}{i\lambda\Delta z}\int_{-D/2}^{D/2}\int_{-D/2}^{D/2} e^{i\frac{k}{2\Delta z}\left[(x_1-x_2)^2+(y_1-y_2)^2\right]}\,dx_1\,dy_1. \tag{1.59}$$

The details of the steps involved in solving this integral are given in Fourier-optics textbooks like Goodman (Ref. 5). The solution, making use of Fresnel sine and cosine integrals is given by

$$\begin{aligned} U(x_2,y_2) = \frac{e^{ik\Delta z}}{2i}&\left\{\left[C(\alpha_2)-C(\alpha_1)\right]+i\left[S(\alpha_2)-S(\alpha_1)\right]\right\} \\ &\times\left\{\left[C(\beta_2)-C(\beta_1)\right]+i\left[S(\beta_2)-S(\beta_1)\right]\right\}, \end{aligned} \tag{1.60}$$

where

$$\alpha_1 = -\sqrt{\frac{2}{\lambda\Delta z}}\left(\frac{D}{2}+x_2\right) \tag{1.61}$$

$$\alpha_2 = \sqrt{\frac{2}{\lambda\Delta z}}\left(\frac{D}{2}-x_2\right) \tag{1.62}$$

$$\beta_1 = -\sqrt{\frac{2}{\lambda\Delta z}}\left(\frac{D}{2}+y_2\right) \tag{1.63}$$

$$\beta_2 = \sqrt{\frac{2}{\lambda\Delta z}}\left(\frac{D}{2}-y_2\right). \tag{1.64}$$

In Eq. (1.60), $S(x)$ and $C(x)$ are the Fresnel sine and cosine integrals given by

$$S(x) = \int_0^x \sin\left(\frac{\pi t^2}{2}\right) dt \tag{1.65}$$

$$C(x) = \int_0^x \cos\left(\frac{\pi t^2}{2}\right) dt, \tag{1.66}$$

respectively. MATLAB code for evaluating this solution is given in Appendix B.

Numerically evaluating the Fresnel diffraction integral with accurate results poses some interesting challenges. These challenges are due to using discrete samples on a finite-sized grid, which is required to evaluate this integral on a digital computer. Basic analysis of these issues is discussed in Ch. 2, which actually focuses on Fourier transforms because they arise so often in scalar diffraction theory. In fact, Eq. (1.57) can be written in terms of a Fourier transform, which is desirable because discrete Fourier transforms can be computed with great efficiency.

After Ch. 2 discusses discrete Fourier transforms, Ch. 3 discusses several basic computations that can be written in terms of Fourier transforms. Chapter 4 presents this book's first application of discrete Fourier transforms to optics by studying situations with very far propagation distances through free space and situations with lenses. These conditions allow simplifications to Eq. (1.57). For example, when we assume that the propagation distance Δz is very far, we can approximate the quadratic phase factor in Eq. (1.57) as being flat. Specifically, we must have $\Delta z > 2D^2/\lambda$, where D is the maximum spatial extent of the source-plane field[5]. This is the Fraunhofer approximation, which leads to the Fraunhofer diffraction integral

$$U(x_2, y_2) = \frac{e^{ik\Delta z} e^{i\frac{k}{2\Delta z}\left(x_2^2+y_2^2\right)}}{i\lambda\Delta z} \int_{-\infty}^{\infty}\int_{-\infty}^{\infty} U(x_1, y_1)\, e^{-i\frac{k}{\Delta z}(x_1 x_2 + y_1 y_2)}\, dx_1 dy_1. \tag{1.67}$$

As an example of a Fraunhofer diffraction pattern, consider a planar wave passing through a two-slit aperture in an opaque screen. With two rectangular slits, the field just after the screen is

$$U(x_1, y_1) = \left[\mathrm{rect}\left(\frac{x_1 - \Delta x/2}{D_x}\right) + \mathrm{rect}\left(\frac{x_1 + \Delta x/2}{D_x}\right)\right] \mathrm{rect}\left(\frac{y_1}{D_y}\right), \tag{1.68}$$

where the slits are D_x wide in the x_1 direction and D_y wide in the y_1 direction and $\Delta x > D_x$ is the distance between the slits' centers. The resulting observation-plane field is

$$U(x_2, y_2) = \frac{e^{ik\Delta z} e^{i\frac{k}{2\Delta z}(x_2^2+y_2^2)}}{i\lambda\Delta z} \int_{-\infty}^{\infty}\int_{-\infty}^{\infty} \left[\text{rect}\left(\frac{x_1 - \Delta x/2}{D_x}\right) + \text{rect}\left(\frac{x_1 + \Delta x/2}{D_x}\right)\right]$$

$$\times \text{rect}\left(\frac{y_1}{D_y}\right) e^{-i\frac{k}{\Delta z}(x_1 x_2 + y_1 y_2)}\, dx_1 dy_1 \tag{1.69}$$

$$= \frac{e^{ik\Delta z} e^{i\frac{k}{2\Delta z}(x_2^2+y_2^2)}}{i\lambda\Delta z} \left[\int_{-(\Delta x+D_x)/2}^{(-\Delta x+D_x)/2} e^{-i\frac{k}{\Delta z}x_1 x_2}\, dx_1 + \int_{(\Delta x-D_x)/2}^{(\Delta x+D_x)/2} e^{-i\frac{k}{\Delta z}x_1 x_2}\, dx_1\right]$$

$$\times \int_{-D_y/2}^{D_y/2} e^{-i\frac{k}{\Delta z}y_1 y_2}\, dy_1 \tag{1.70}$$

$$= e^{ik\Delta z} e^{i\frac{k}{2\Delta z}(x_2^2+y_2^2)} \frac{2D_x D_y}{i\lambda\Delta z} \cos\left(\frac{\pi\Delta x\, x_2}{\lambda\Delta z}\right) \text{sinc}\left(\frac{D_x x_2}{\lambda\Delta z}\right) \text{sinc}\left(\frac{D_y y_2}{\lambda\Delta z}\right). \tag{1.71}$$

While fully coherent illumination was used here, two-slit apertures like this are useful for studying partially coherent sources.[6]

Further problems involving Fraunhofer (Ch. 4) and Fresnel (Chs. 6–8) diffraction are studied and simulated later in the book.

1.4 Problems

1. Using Maxwell's equations, show that

$$\mathbf{E} = -\frac{c^2}{2\pi\nu}\mathbf{k} \times \mathbf{B} \tag{1.72}$$

for a planar wave propagating through vacuum.

2. Using Maxwell's equations, show that

$$\mathbf{B} = \frac{1}{2\pi\nu}\mathbf{k} \times \mathbf{E} \tag{1.73}$$

for a planar wave propagating through vacuum.

3. A diverging spherical wave is the result of a Dirac delta-function source. Show that when the source field $U(\mathbf{r}_1) = \delta(\mathbf{r}_1)$ is substituted into the Fresnel diffraction integral, the observation-plane field $U(\mathbf{r}_2)$ is a paraxial spherical wave.

4. Write the scalar wave equation in spherical coordinates and show that the spherical wave is a solution.

5. Suppose that a spherical wave given by

$$U(\mathbf{r}_1) = A\frac{e^{ikR_1}}{R_1}e^{i\frac{k}{2R_1}\left[(x-x_c)^2+(y-y_c)^2\right]} \tag{1.74}$$

   is the optical field in the source plane. Substitute this into Eq. (1.57) to compute the optical field $U(\mathbf{r}_2)$ in the observation plane.

6. Suppose that a monochromatic, uniform-amplitude planar wave has passed through an annular circular aperture, and immediately after the aperture, the field is given by

$$U(\mathbf{r}_1) = \text{circ}\left(\frac{2r_1}{D_{out}}\right) - \text{circ}\left(\frac{2r_1}{D_{in}}\right), \tag{1.75}$$

   where $D_{out} > D_{in}$. Use the Fraunhofer diffraction integral to compute the observation-plane field (far away).

Chapter 2
Digital Fourier Transforms

As discussed in Ch. 1, scalar diffraction theory is the physical basis of wave-optics simulations. A result of this theory is that propagation of electromagnetic waves through vacuum may be treated as a linear system. For monochromatic waves, the vector magnitude of the electric field in the observation plane of a system is the convolution of the vector magnitude of the electric field in the source plane and the free-space impulse response.[5] Consequently, the tools of linear-systems theory and Fourier analysis are indispensable for studying wave optics. These topics are discussed in Ch. 4 and beyond. In those chapters, discrete Fourier transforms are applied to obtain computationally efficient algorithms for the simulations. First, the basic computational algorithms must be discussed.

As in many areas of science and engineering, most problems encountered while researching complex optical systems are analytically intractable. Consequently, most calculations regarding the inner workings and performance of optical systems are performed by numerical simulation on computers. Fortunately, sampling theory and discrete-Fourier-transform (DFT) theory provide many important lessons for optics researchers who perform such simulations. With due consideration to the limitations imposed by performing computations on sampled functions, there is much to be gained from numerical simulation of optical-wave propagation.

2.1 Basics of Digital Fourier Transforms

This section covers the basics of computing DFTs that match the corresponding analytic results. This includes proper scaling, correct use of spatial and spatial-frequency coordinates, and use of DFT software.

2.1.1 Fourier transforms: from analytic to numerical

There are a few common conventions for defining the FT operation and its inverse. This book defines the continuous FT $G(f_x)$ of a spatial function $g(x)$ and its inverse as

$$G(f_x) = \mathcal{F}\{g(x)\} = \int_{-\infty}^{\infty} g(x)\, e^{-i2\pi f_x x}\, dx \tag{2.1}$$

$$g(x) = \mathcal{F}^{-1}\{G(f_x)\} = \int_{-\infty}^{\infty} G(f_x)\, e^{i2\pi f_x x}\, df_x, \tag{2.2}$$

where x is the spatial variable, and f_x is the spatial-frequency variable. The first step to discretize the FT is writing the integral as a Riemann sum:

$$\begin{aligned} G(f_{xm}) &= \mathcal{F}\{g(x_n)\} \\ &= \sum_{n=-\infty}^{\infty} g(x_n)\, e^{-i2\pi f_{xm} x_n}(x_{n+1} - x_n), \quad m = -\infty, \ldots \infty, \end{aligned} \tag{2.3}$$

where n and m are integers. Computer calculations can only work with a finite number of samples N, and this book discusses only even N for reasons that are discussed later. Further, typical DFT software requires a fixed sampling interval. The sampling interval is δ, and so $x_n = n\delta$. Then, the frequency domain interval is $\delta_f = 1/(N\delta)$ such that $f_{xm} = m\delta_f = m/(N\delta)$. Eq. (2.3) becomes

$$\begin{aligned} G\left(\frac{m}{N\delta}\right) &= \mathcal{F}\{g(n\delta)\} \\ &= \delta \sum_{n=-N/2}^{N/2-1} g(n\delta)\, e^{-i2\pi mn/N}, \qquad m = -N/2, 1 - N/2, \ldots N/2 - 1. \end{aligned} \tag{2.4}$$

The last step is to format the samples for the DFT software. Such software is available for many programming languages. Examples in this book use the MATLAB scripting language, which has DFT routines in its core function library.[7] Other programming languages such as C, C++, FORTRAN, and Java do not have DFT routines in their core libraries, but DFT algorithms are described in many books,[8] and DFT software is readily available from third-party suppliers.[9–11] MATLAB uses positive indices (also called one-based indexing). To account for only positive indices, the order of the spatial samples inside the sum must be rearranged such that

$$g_{n'} = \begin{cases} g\left[\left(n' + \frac{N}{2}\right)\delta\right] & \text{for} \quad n' = 1, 2, \ldots \frac{N}{2} + 1 \\ g\left[(n' - N - 2)\,\delta\right] & \text{for} \quad n' = \frac{N}{2} + 2, \frac{N}{2} + 3, \ldots N. \end{cases} \tag{2.5}$$

For a one-dimensional DFT, this amounts to circularly shifting the samples in the spatial domain so that the origin corresponds to the first sample, as illustrated in Fig. 2.1.

The reordering of spatial samples means that the samples in the spatial-frequency domain end up out of order, too. We denote the new index in the spatial-frequency domain as m', which finally leads to the form of the DFT equation:

$$G_{m'} = \delta \sum_{n'=1}^{N} g_{n'} e^{-i2\pi(m'-1)(n'-1)/N}, \qquad m' = 1, 2, \ldots N. \tag{2.6}$$

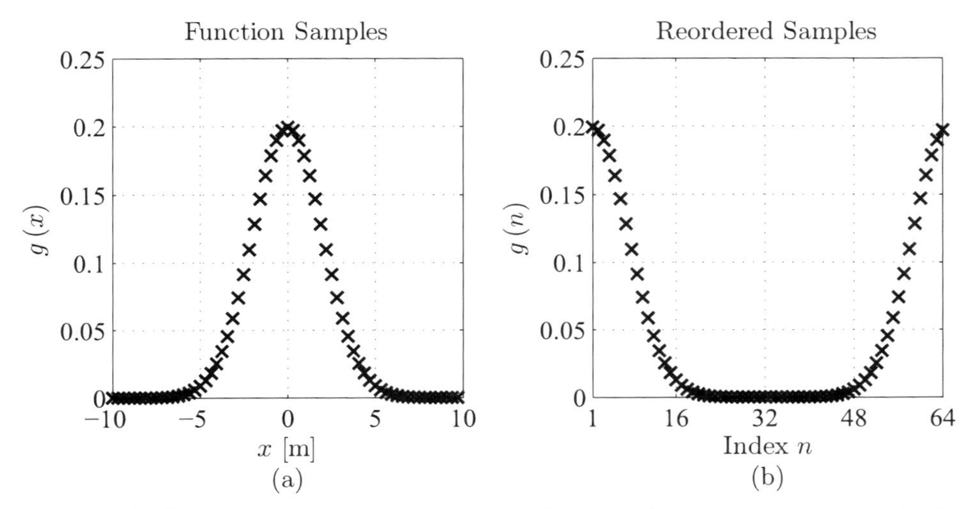

Figure 2.1 An illustration of reordering samples in the spatial domain in preparation for the DFT. Plot (a) shows a Gaussian function in the spatial domain. Plot (b) shows the samples of plot (a) reordered. The reordering essentially circularly shifts the samples so that the origin is at the first element.

MATLAB's DFT software computes everything in Eq. (2.6) except for multiplying by δ, as is typical. That is left to the user.

2.1.2 Inverse Fourier transforms: from analytic to numerical

Discrete IFTs (DIFTs) operate very similarly to DFTs. As before, the first step is to write the integral in Eq. (2.2) as a Riemann sum:

$$\begin{aligned} g(x_n) &= \mathcal{F}^{-1}\{G(f_{xm})\} \\ &= \sum_{m=-\infty}^{\infty} G(f_{xm})\, e^{i2\pi f_{xm} x_n} (f_{x,m+1} - f_{x,m}), \qquad n = -\infty, \ldots \infty. \end{aligned} \tag{2.7}$$

Again, with a finite number of samples N and uniform sample spacing $\delta_f = 1/(N\delta)$ in the frequency domain, the sum becomes

$$\begin{aligned} g(n\delta) &= \mathcal{F}^{-1}\{G(f_{xm})\} \\ &= \delta_f \sum_{m=-N/2}^{N/2-1} G\left(\frac{m}{N\delta}\right) e^{i2\pi mn/N}, \qquad n = -N/2, 1 - N/2, \ldots N/2 - 1. \end{aligned} \tag{2.8}$$

Then, the use of positive indices results in reordering of the samples similar to what happens in the forward DFT. The result is

$$g_{n'} = \frac{1}{N\delta} \sum_{m'=1}^{N} G_{m'} e^{i2\pi(m'-1)(n'-1)/N}, \qquad n' = 1, 2, \ldots N. \tag{2.9}$$

Listing 2.1 Code for performing a DFT in MATLAB.

```
1 function G = ft(g, delta)
2 % function G = ft(g, delta)
3     G = fftshift(fft(fftshift(g))) * delta;
```

Listing 2.2 Code for performing a DIFT in MATLAB.

```
1 function g = ift(G, delta_f)
2 % function g = ift(G, delta_f)
3     g = ifftshift(ifft(ifftshift(G))) ...
4         * length(G) * delta_f;
```

DFT software typically computes everything in Eq. (2.9) except for multiplying by δ^{-1}.

2.1.3 Performing discrete Fourier transforms in software

MATLAB is one of many software applications that provide DFT functionality.[9–11] Specifically, it includes the functions `fft` and `ifft` for performing one-dimensional DFTs using the fast Fourier-transform (FFT) algorithm. The FFT algorithm works only for values of N that are an integer power of two. Now, this is common practice, but using powers of two is not entirely necessary anymore because of sophisticated DFT software like FFTW (Fastest Fourier Transform in the West).[9] Computational efficiency for DFTs is maximized when N is a power of two, although depending on the value, other lengths can be computed nearly as fast. In any case, we restrict our discussions to only even N, as previously mentioned. Listings 2.1 and 2.2 give functions that compute a properly scaled FT and IFT, making use of `fft` and `ifft`. Listing 2.1 evaluates Eq. (2.6) including the reordering in both domains using the function `fftshift`. Listing 2.2 evaluates Eq. (2.9) including the reordering in both domains using the function `ifftshift`.

Listings 2.3 and 2.4 give examples of computing properly scaled DFTs, making use of `ft` and `ift`, and Figs. 2.2 and 2.3 illustrate the results. In the first example, both the spatial function and its spectrum are real and even. In the second example, the spatial function is a shifted version of that from the first example. The result of the shift is a non-zero phase in the spectrum.

Figure 2.2 shows that the DFT values for a Gaussian function match the analytic FT values closely. The most notable departure is at $f_x = 0$. However, if the original function were to be synthesized from the DFT values shown in Fig. 2.2, any error at $f_x = 0$ would only affect the mean value of synthesized function, not its structure.

Figure 2.3 shows that the DFT values for a shifted Gaussian function match the

Listing 2.3 MATLAB example of performing a DFT with comparison to the analytic FT. The spatial function is real and even.

```
1  % example_ft_gaussian.m
2
3  % function values to be used in DFT
4  L = 5;          % spatial extent of the grid
5  N = 32;         % number of samples
6  delta = L / N;  % sample spacing
7  x = (-N/2 : N/2-1) * delta;
8  f = (-N/2 : N/2-1) / (N*delta);
9  a = 1;
10 % sampled function & its DFT
11 g_samp = exp(-pi*a*x.^2);  % function samples
12 g_dft = ft(g_samp, delta);  % DFT
13 % analytic function & its continuous FT
14 M = 1024;
15 x_cont = linspace(x(1), x(end), M);
16 f_cont = linspace(f(1), f(end), M);
17 g_cont = exp(-pi*a*x_cont.^2);
18 g_ft_cont = exp(-pi*f_cont.^2/a)/a;
```

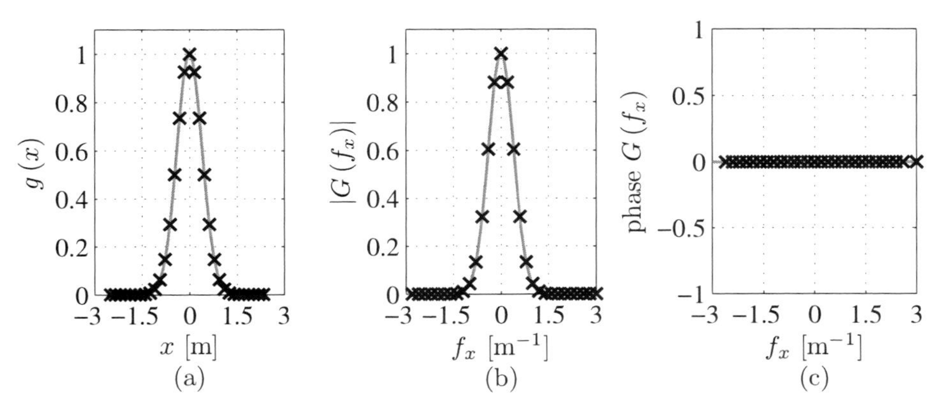

Figure 2.2 A Gaussian function and its properly scaled DFT plotted with its analytic counterpart.

Listing 2.4 MATLAB example of performing a DFT with comparison to the analytic FT. The spatial function is real but asymmetric.

```
1  % example_ft_gaussian_shift.m
2
3  L = 10;        % spatial extent of the grid
4  N = 64;        % number of samples
5  delta = L / N; % sample spacing
6  x = (-N/2 : N/2-1) * delta;
7  x0 = 5*delta;
8  f = (-N/2 : N/2-1) / (N*delta);
9  a = 1;
10 % sampled function & its DFT
11 g_samp = exp(-pi*a*(x-x0).^2); % function samples
12 g_dft = ft(g_samp, delta); % DFT
13 % analytic function & its continuous FT
14 M = 1024;
15 x_cont = linspace(x(1), x(end), M);
16 f_cont = linspace(f(1), f(end), M);
17 g_cont = exp(-pi*a*(x_cont-x0).^2);
18 g_ft_cont = exp(-i*2*pi*x0*f_cont) ...
19     .* exp(-pi*f_cont.^2/a)/a;
```

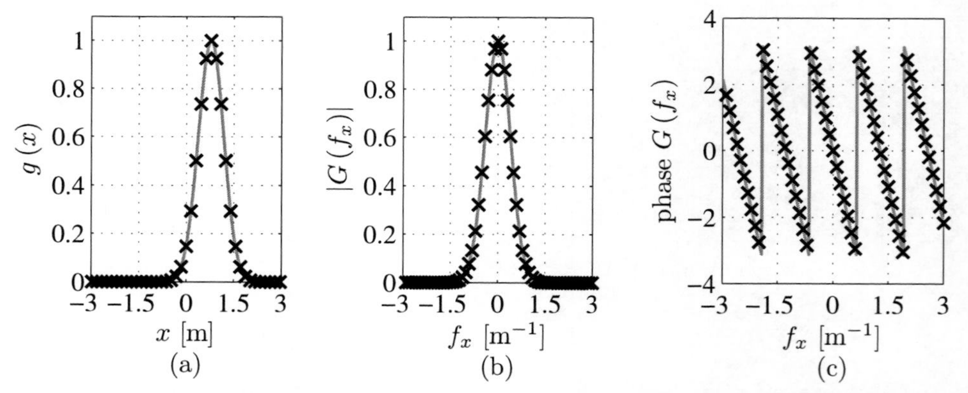

Figure 2.3 A shifted Gaussian function and its properly scaled DFT plotted with its analytic counterpart. Plot (a) shows the spatial function and its sample values. Plot (b) shows the modulus of the analytic FT and the modulus of the DFT. Plot (c) shows the analytic phase of the FT and the phase of the DFT.

analytic FT values closely. The spatial shift moved the Gaussian pulse toward one edge of the grid. As a result, the grid had to be extended to twice the size shown in Fig. 2.2 by doubling the number of samples. Without the increased number of samples, the phase in the spatial-frequency domain would match the analytic result only in the center of the spectrum.

2.2 Sampling Pure-Frequency Functions

A very important issue in achieving accurate results with FTs and FT-based calculations is determining the necessary grid spacing δ and number of grid points N. This is an important distinction between Figs. 2.2 and 2.3. The highest significant frequency in the shifted Gaussian signal is higher than that in the centered Gaussian. Accordingly, the shifted Gaussian requires more samples to adequately represent its spectrum. The reasons for this requirement are discussed in this section.

The Whittaker-Shannon sampling theorem states that a bandlimited signal having no spectral components above f_{max} can be uniquely determined by values sampled at uniform intervals of $\delta_c = 1/(2f_{max})$.[5,12] The Nyquist sampling frequency is defined as $f_c = 1/\delta_c = 2f_{max}$. The requirement for sampling frequencies higher than f_c is called the Nyquist sampling criterion. Essentially, this means that there must be at least two samples per period for the highest frequency component of the signal. If the sample spacing is larger than δ_c, it may not be possible to reconstruct each frequency component uniquely. This can be a problem for DFTs.

The simplest way to illustrate sampling effects is with pure sinusoidal signals. The following discussion can be extended to any Fourier-transformable signal by applying the Fourier integral representation. This section uses signals of the form

$$g(x) = \cos(2\pi f_0 x) \tag{2.10}$$

to illustrate some aspects of sampling related to this theorem. In this type of signal, the frequency is f_0 and the period is $T = 1/f_0$. The required grid spacing is $\delta_c = 1/(2f_0)$, corresponding to two samples per period.

Figure 2.4 shows such a sinusoidal signal. This particular signal, shown by the solid gray line, has a frequency of $6\ \mathrm{m}^{-1}$. Samples of the signal, separated by $\delta_0 = 1/12\ \mathrm{m} = 0.0833\ \mathrm{m}$, are shown in the black $\times$'s. The samples are located at all of the peaks and troughs of the signal. Now, if we were given these samples without knowledge of the signal from which they were drawn, could we uniquely identify the signal? Actually, there are are many other sinusoidal signals that could have produced these samples. For example, $\cos(4\pi f_0 x)$ could produce the samples shown; however, there is no frequency lower than f_0 that could have produced these samples. Further, the only signal satisfying the Nyquist criterion is f_0. Now, we realize that if we are given the samples and the fact that they satisfied the Nyquist criterion, we could certainly identify the signal uniquely.

As a counter-example, we consider sinusoidal signals that are sampled on grids that do not satisfy the Nyquist criterion. Figure 2.5 shows two such signals. In

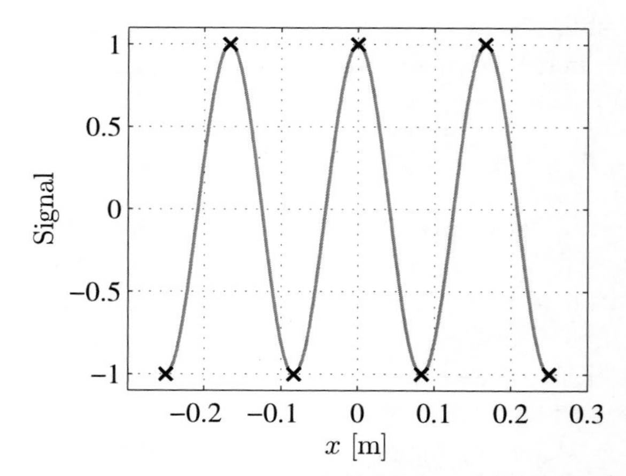

Figure 2.4 Example of a sinusoidal signal (gray line) that is properly sampled. There is no lower frequency that could produce the samples shown.

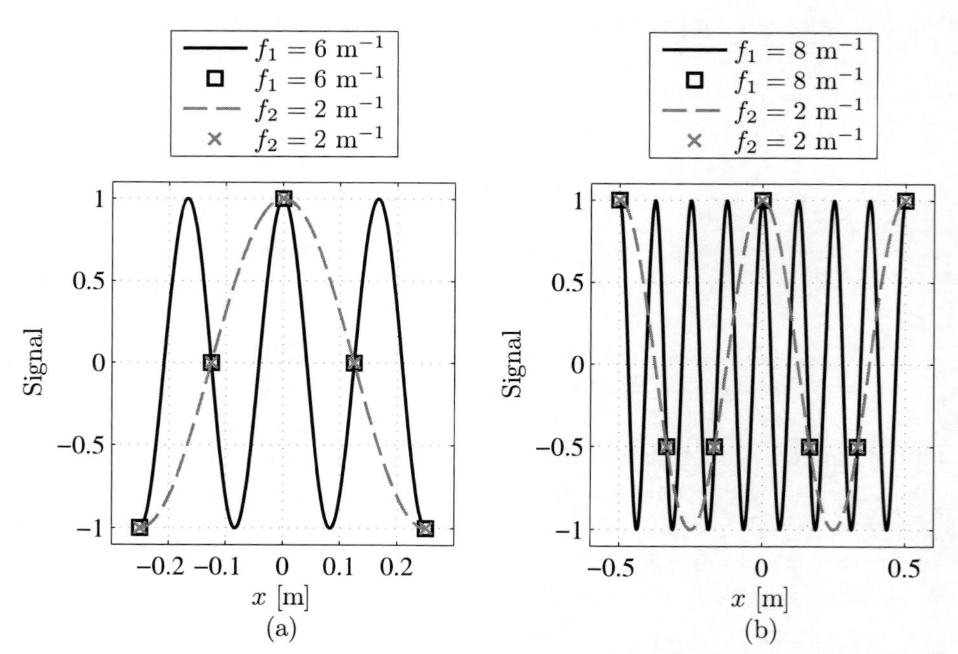

Figure 2.5 Example of a sinusoidal signal (gray line) that is sampled too coarsely. Samples taken from both frequencies are identical.

plot (a), the solid black line shows a cosine signal with frequency $f_1 = 6\ \text{m}^{-1}$. Properly sampling this signal would require a sample spacing of $1/12$ m$= 0.0833$ m. The black squares show samples of this signal that are separated by $\delta = 1/8$ m$=$ 0.125 m. Now, let us consider the other signal in plot (a). The gray dashed line shows a signal with frequency $f_2 = 2\ \text{m}^{-1}$, and the gray $\times$s show its samples. The samples from the two different frequencies are identical! In the previous example of a properly-sampled function, only frequencies that are multiples of the original, in this case f_1, could produce the given samples. None of those harmonics would be properly sampled, though. Now, when the signal is undersampled, there is at least one lower (and properly-sampled) frequency that could produce the given samples. If we were given these samples and someone asked us to identify the signal's frequency, and we answered with a properly-sampled signal (satisfying the Nyquist criterion), like $2\ \text{m}^{-1}$, we would be incorrect.

This is not a rare occurrence; plot (b) shows another undersampled example with $f_1 = 8\ \text{m}^{-1}$ sampled with a grid spacing of $1/6$ m$= 0.167$ m. Again, the gray dashed line shows a signal with frequency $f_2 = 2\ \text{m}^{-1}$, and its samples shown in gray $\times$s are identical to those taken from the higher frequency. When the grid spacing is too coarse, the improperly-sampled, high-frequency sinusoids appear as properly-sampled, lower frequencies. This effect is called aliasing.

Returning to other signals that can be written as a sum or integral of sinusoids, we need to know the highest frequency component and then compute the grid spacing from there. If the highest frequency is properly sampled, so are all of the lower frequencies. This seems like a simple solution, but there are many examples in this book that are not so straightforward, and even cases in which we can (and probably should) relax this constraint. The next section gives a more detailed treatment.

2.3 Discrete vs. Continuous Fourier Transforms

DFT pairs differ from their continuous counterparts in three important ways:

- spatial domain sampling,
- a finite spatial grid,
- and spatial-frequency-domain sampling.

These three properties result in three distortions to continuous FT pairs when they are computed discretely:

- aliasing in the spatial-frequency domain,
- rippling and smearing in the spatial-frequency domain,
- and virtual periodic replication in the spatial domain.

These effects are illustrated more formally here in a development that closely follows the approach of Brigham.[8] Let a known FT pair be

$$g(x) \Leftrightarrow G(f_x), \tag{2.11}$$

and let the sampled versions of these functions be

$$\widetilde{g}(x) \Leftrightarrow \widetilde{G}(f_x), \tag{2.12}$$

respectively. The next few equations develop the sampled FT pair. Figure 2.6 shows the graphical development. The figure uses

$$g(x) = \exp(-a|x|) \tag{2.13}$$

$$G(f_x) = \frac{1}{a}\frac{2}{1+(2\pi f_x/a)^2} \tag{2.14}$$

as the example FT pair to illustrate the effects of discretization. This is for illustration purposes; the effects would be the same for any other FT pair. Plots of Eqs. (2.13) and (2.14) are shown in Figs. 2.6 (a) and (b) for $a = 10\,\mathrm{m}^{-1}$. The peak value of the spectrum is 0.2.

To begin accounting for discretization, $g(x)$ is sampled by multiplication with a comb function with spacing δ. Multiplication in the spatial domain is equivalent to convolution in the spatial-frequency domain (for a discussion of convolution, see Ch. 3), which transforms the pair in Eq. (2.11) into

$$g(x)\frac{1}{\delta}\,\mathrm{comb}\left(\frac{x}{\delta}\right) \Leftrightarrow G(f_x) \otimes \mathrm{comb}(\delta f_x). \tag{2.15}$$

Figures 2.6(c) and (d) show the impact of sampling in the spatial domain for $\delta = 0.0375$ m. This results in periodic replication in the spatial-frequency domain. This is visible in the tails of the frequency spectrum that lift up at large positive and negative frequencies. That is an artifact that is not present in the analytic spectrum shown in Fig. 2.6(b).

Next, representing $g(x)$ on a grid of finite size L changes the pair into

$$g(x)\frac{1}{\delta}\,\mathrm{comb}\left(\frac{x}{\delta}\right)\mathrm{rect}\left(\frac{x}{L}\right) \Leftrightarrow G(f_x) \otimes \mathrm{comb}(\delta f_x) \otimes [L\ \mathrm{sinc}(L f_x)]. \tag{2.16}$$

Figures 2.6(e) and (f) show the impact of the finite sample width, $L = 0.6\,\mathrm{m}$. In the spatial domain, the tails of $g(x)$ are lost. In the spatial-frequency domain, the spectrum is multiplied by L and convolved with a sinc function, which causes rippling and smearing.

Finally, the result of the DFT is an array of the sampled values of $G(f_x)$. This makes one final modification to the FT pair so that

$$\widetilde{g}(x) = \left[g(x)\frac{1}{\delta}\,\mathrm{comb}\left(\frac{x}{\delta}\right)\mathrm{rect}\left(\frac{x}{L}\right)\right] \otimes \left[\frac{1}{L}\,\mathrm{comb}\left(\frac{x}{L}\right)\right] \tag{2.17}$$

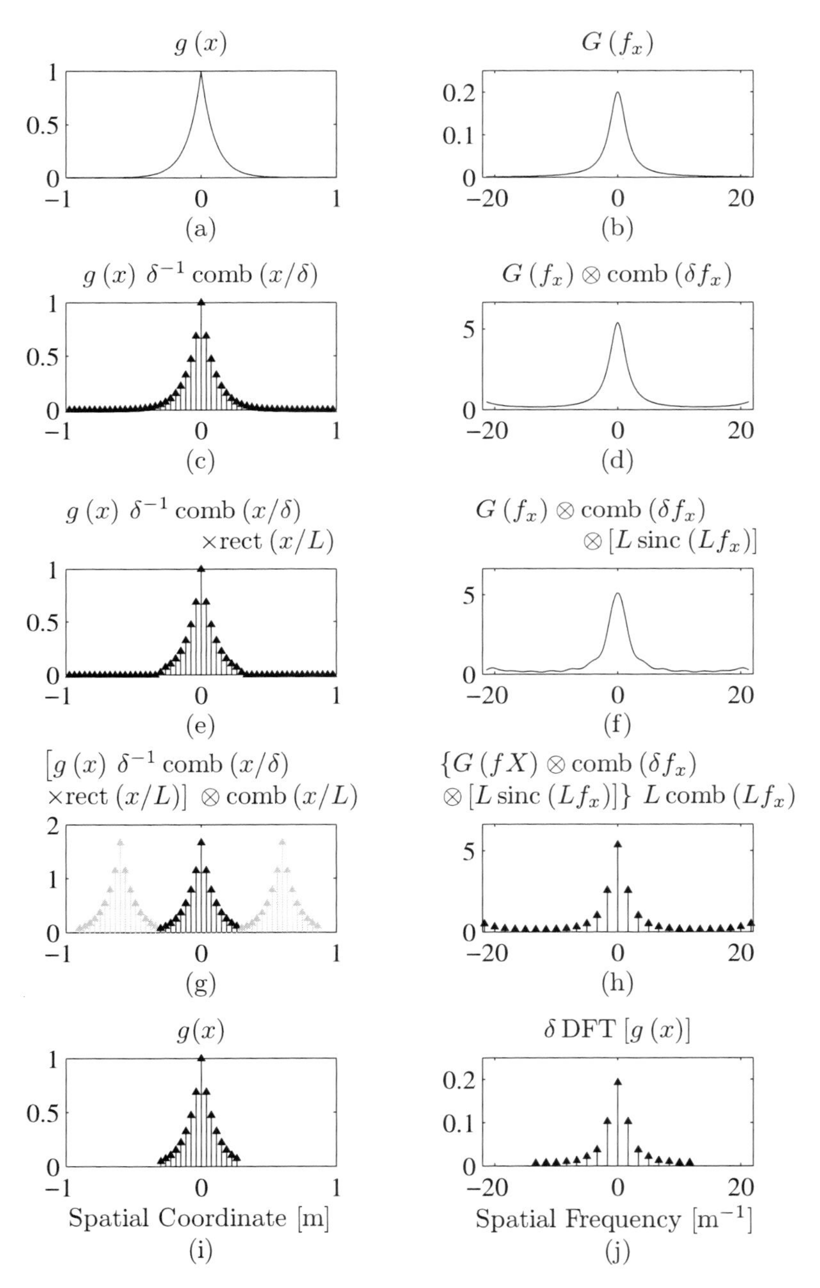

Figure 2.6 Graphical development of the DFT from the analytic FT.

$$\widetilde{G}(f_x) = [G(f_x) \otimes \text{comb}(\delta f_x) \otimes L \ \text{sinc}(L f_x)] \times \text{comb}(L f_x). \tag{2.18}$$

The impact of sampling the spatial-frequency domain is shown in Fig. 2.6(g). The result is virtual periodic replication in the spatial domain. The term 'virtual' is used because there are actually no samples in the periodically replicated region.

Figures 2.6(i) and (j) show the final DFT pair. Plot (i) shows only the samples from the spatial domain that input to the DFT algorithm, and Plot (j) shows the output from the `ft` function.

To provide a clarification, the reader should note that one effect has not been discussed yet. Figure 2.6(h) shows a frequency function that still has an infinite number of samples. One would logically expect that we should go a step further and account for the finite number of samples with multiplication by a rect function in the frequency domain. This would imply that the spatial-domain function is rippled and broadened by convolution with a sinc function. However, we are considering a forward FT so that we start with the black samples shown in Fig. 2.6(g), which begin undistorted by any such convolution. Now, if we were to consider a discrete IFT, we could simply treat plots (a), (c), (e), (g), and (i) as the frequency-domain function. The IFT differs from the forward FT by only a sign in the exponential, which does not affect these distortions. Consequently, if we start with an undistorted frequency-domain function and perform a discrete IFT, the spatial-domain function would be periodically replicated, rippled, and sampled like in plots (b), (d), (f), (h), and (j).

2.4 Alleviating Effects of Discretization

When we want to use a DFT to approximate a continuous FT $G(f_x)$ of a known function $g(x)$, the FT pair that is actually used is $\widetilde{g}(x)$ and $\widetilde{G}(f_x)$ as given by Eqs. (2.17) and (2.18). The result $\widetilde{G}(f_x)$ of the DFT is a sampled, rippled, and aliased version of the desired analytic result. These effects may be reduced, but usually not eliminated. The rippling may be reduced by increasing the spatial grid size L, and the aliasing may be reduced by decreasing the spatial grid spacing δ.

Figures 2.7, 2.8, and 2.9 illustrate the results of various attempts to limit rippling and aliasing (as compared to Fig. 2.6). In producing Fig. 2.7, a larger grid has been used by increasing δ while keeping N the same. As a result, the factor $L\,\text{sinc}(L f_x)$ became narrower, thereby reducing the rippling. This can be seen by comparing Fig. 2.7(f) to Fig. 2.6(f). Unfortunately, increasing δ means that the factor $\text{comb}(\delta f_x)$ now has a narrower spacing, leading to increased aliasing, which is visible in Fig. 2.7(d). Conversely, in producing Fig. 2.8, more samples have been used so that N has increased, δ has decreased, and L remains the same. This approach reduces the aliasing by spreading out the $\text{comb}(\delta f_x)$ factor, but without improving the rippling. The reduced aliasing is evident in Fig. 2.8(d), and the unchanged rippling is visible in Fig. 2.8(f). Finally, in learning a lesson from Figs. 2.7 and 2.8, smaller δ and larger L were used in producing Fig. 2.9. This approach

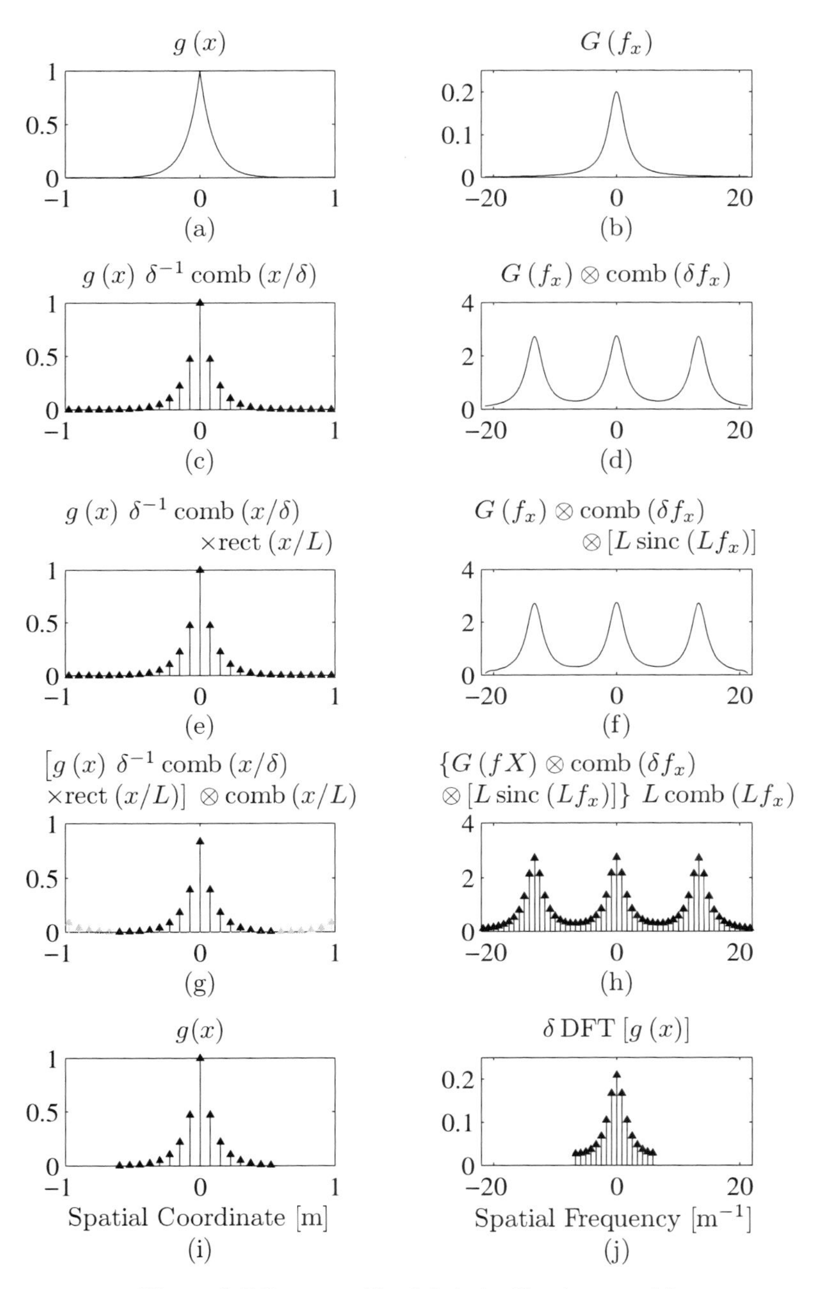

Figure 2.7 Same as Fig. 2.6, but with a larger grid.

reduces aliasing and rippling at the same time, which is clearly the best approach. The drawbacks are the additional memory and computations required.

Unlike the graphical example above, some functions are strictly bandlimited. This means that the function $g(x)$ that we want to transform has a maximum fre-

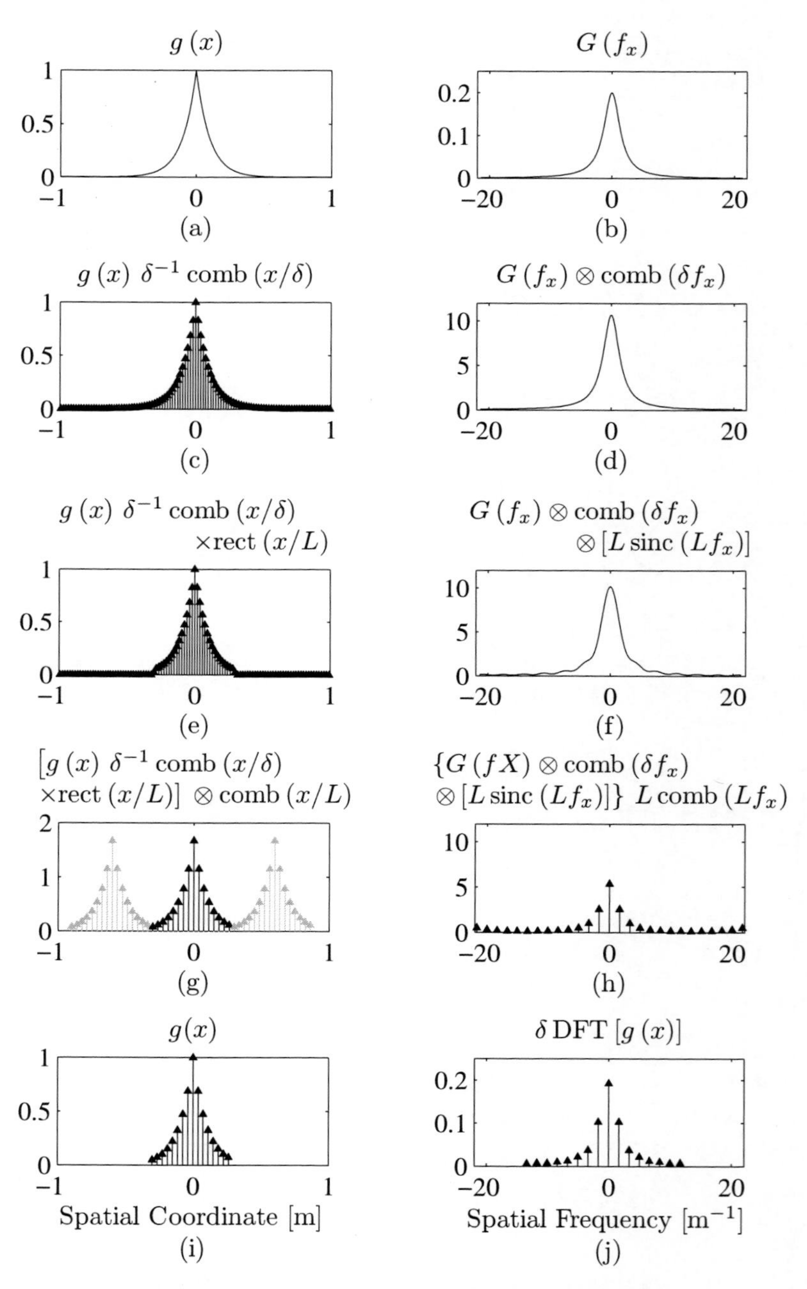

Figure 2.8 Same as Fig. 2.6, but with more samples.

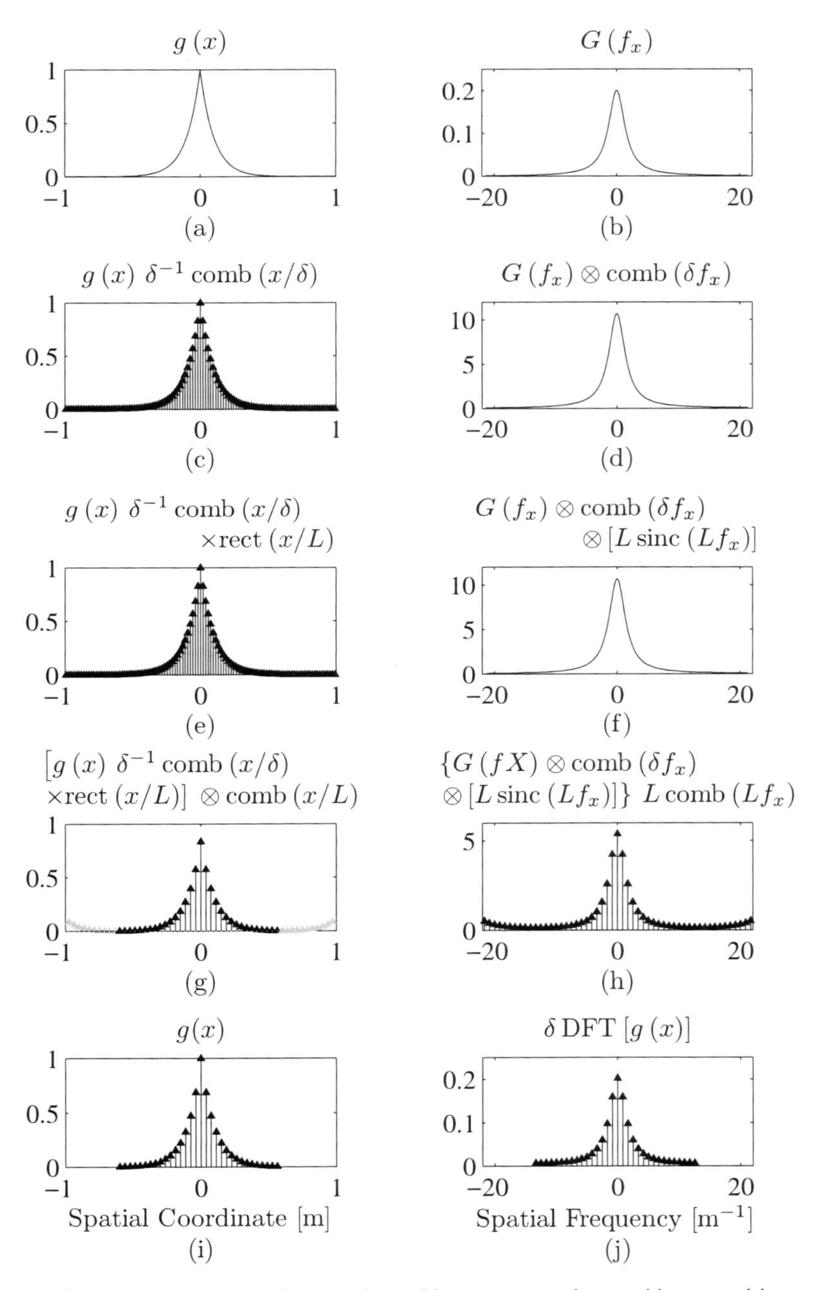

Figure 2.9 Same as Fig. 2.6, but with more samples and larger grid.

quency $f_{x,max}$ such that

$$G(f_x) = 0 \text{ for } |f_x| > f_{x,max} \tag{2.19}$$

for some finite spatial frequency $f_{x,max}$. This frequency is called the bandwidth of $g(x)$. As discussed in Sec. 2.2, if we sample this continuous function so that there are two samples for every cycle of the highest frequency component, the continuous function can be reconstructed exactly from its spectrum. This requirement on the grid spacing can be expressed as

$$\delta \leq \frac{1}{2f_{x,max}}. \tag{2.20}$$

This is a very important consideration in the chapters covering Fresnel diffraction. Ch. 7 discusses this in detail.

Like the graphical example, sometimes signals are not strictly bandlimited, but there is a limit to how much bandwidth the user cares about. If he is simulating a system that can only sample at a rate of f_s, then the sampling requirement can be relaxed to

$$\delta \leq \frac{1}{f_s + f_{x,max}}. \tag{2.21}$$

This way, aliasing is present but not in the frequency range that the user cares about. The aliased frequencies wrap around from one edge of the grid to the edge of the other side, only distorting the spectrum at the highest frequencies.

2.5 Three Case Studies in Transforming Signals

In optics, we apply the FT to many types of signals with different types of band limits. This section highlights three different signals and how to compute their DFTs accurately. Computing the spectra of these deterministic signals provides important lessons for later when we want to compute the spectra of unknown and sometimes random signals. The three signals are a sinc, a Gaussian, and a Gaussian $\times$ a quadratic phase. The first of these cases has a "hard" band limit like in Appendix A, while the latter two have "soft" band limits. Each case highlights different sampling considerations that become very important in later chapters.

2.5.1 Sinc signals

The sinc signal used in this book is defined in Appendix A. It is a good example of a signal that is intrinsically bandlimited such that its FT values are identically zero beyond a certain maximum frequency. It has a simple analytic FT given by

$$G(f_x) = \frac{1}{a} \operatorname{rect}\left(\frac{f_x}{a}\right). \tag{2.22}$$

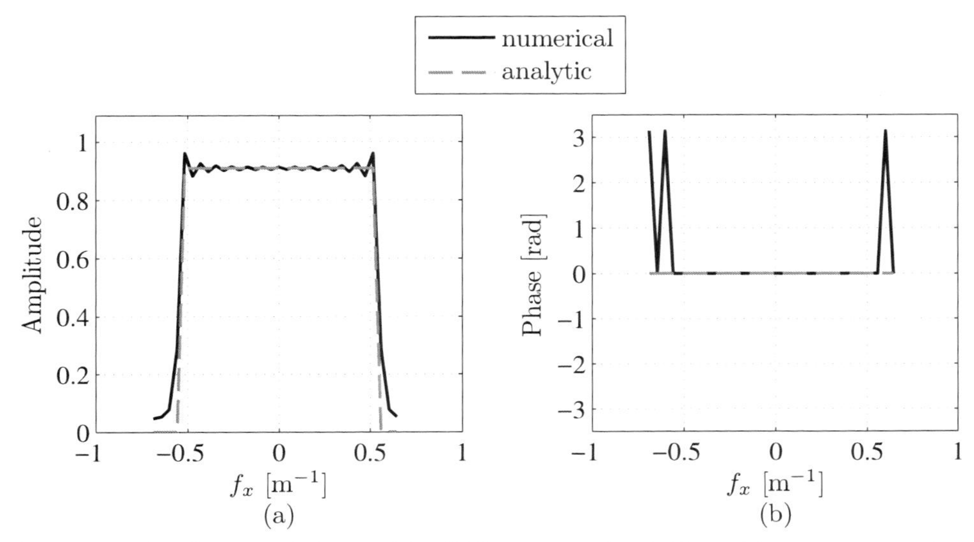

Figure 2.10 Amplitude and phase of the DFT of a sinc signal. The grid spacing was determined by applying the Nyquist criterion.

Because we know the analytic FT of this signal, we know its maximum frequency before computing the DFT. We can then apply the Nyquist criterion to properly sample it before computing the DFT. The maximum frequency in Eq. (2.22) is $a/2$. Applying the Nyquist criterion, we get $\delta \leq 1/(2a/2) = 1/a$. We can try computing the DFT of a sinc signal just below (so that the frequency grid is a little broader than the spectrum) this maximum grid spacing to demonstrate how well it works.

Figure 2.10 shows the DFT of a sinc signal with $a = 1.1$. The solid black line shows the result when the grid spacing is $\delta = 0.85/a$ and $N = 32$. A slight ripple is visible in the amplitude of the DFT shown in plot (a). This is because the spatial grid has not captured the entire spatial extent of the signal. Using more samples (with fixed grid spacing) reduces this ripple. In plot (b), the phase of the DFT at the edge of the frequency grid appears to jump between the correct value, zero, and an incorrect value, π. This is because the DFT values are not exactly zero, which they should be at the edge. They are slightly negative, which is the same as saying that the phase of those points is π radians.

2.5.2 Gaussian signals

The Gaussian signal used in this book is defined by

$$g(x) = \exp\left[-\pi (ax)^2\right]. \tag{2.23}$$

This form of the Gaussian appears in common Fourier-optics textbooks, like Goodman.[5] The Gaussian is a good example of a signal that is very nearly bandlimited,

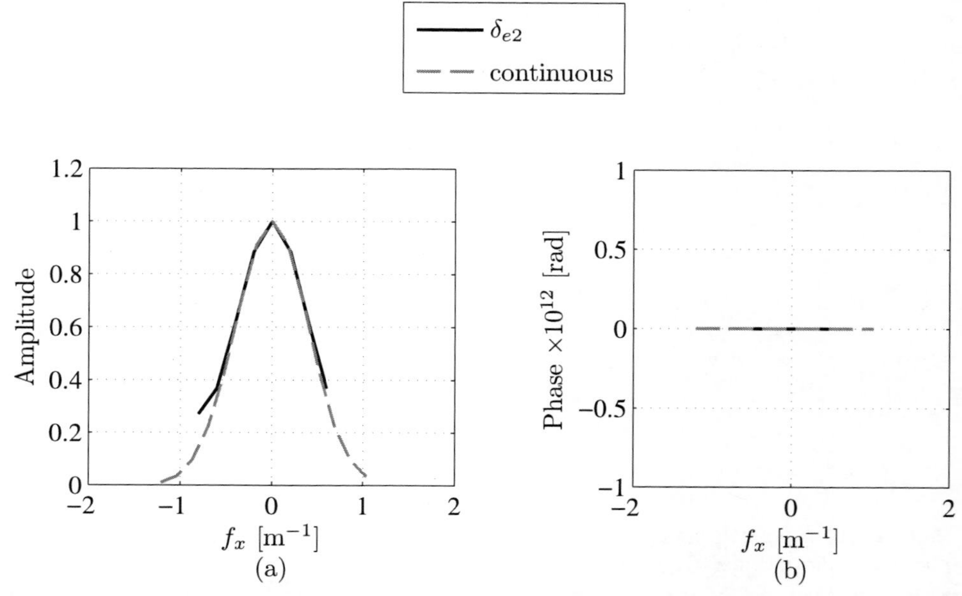

Figure 2.11 Amplitude and phase of the DFT of a Gaussian signal. The grid spacing was determined by applying the Nyquist criterion to the $1/e^2$ frequency.

and it frequently appears in optics because laser beams often have a Gaussian amplitude profile. It has a simple analytic FT given by

$$G(f_x) = \frac{1}{|a|} \exp\left[-\pi (f_x/a)^2\right], \tag{2.24}$$

and its $1/e^2$ frequency is obviously $f_{e2} = a(2/\pi)^{1/2}$. Note that this definition of maximum frequency was arbitrary; we could always choose another definition depending on the situation.

Because we know the analytic FT of this signal, we know its maximum frequency before computing the DFT. We can then apply the Nyquist criterion to properly sample it in advance. Using the $1/e^2$ frequency as $f_{x,max}$, the corresponding maximum grid spacing is

$$\delta_{e2} = \frac{1}{2a}\sqrt{\frac{\pi}{2}}. \tag{2.25}$$

We can try computing the DFT of a Gaussian signal at this maximum grid spacing to see how well it works.

Figure 2.11 shows the DFT of a Gaussian signal with $a = 1$. The solid line shows the result when the grid spacing is δ_{e2}. Aliasing is visible in the left-most sample because a little bit of the spectrum from the right side of the plot, not captured by the samples, wrapped around to the left side. Perhaps the $1/e^2$ is not quite enough to get an accurate DFT.

The value of $1/e^2$ is approximately 0.135; let us try the value p instead, where p has a smaller value, like 0.01. Setting the spectrum equal to $p\times$ its peak value

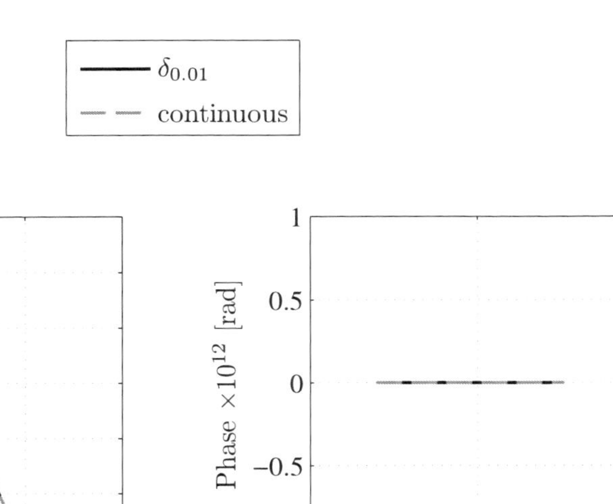

Figure 2.12 Amplitude and phase of the DFT of a Gaussian signal. The grid spacing was determined by applying the Nyquist criterion to the 0.01 frequency.

allows us to solve for the frequency $f_{x,p}$ at this value:

$$p = \exp\left[-\pi\left(f_{x,p}/a\right)^2\right] \tag{2.26}$$

$$f_{x,p} = \left[-\left(\frac{a^2}{\pi}\right)\ln p\right]^{1/2}. \tag{2.27}$$

For example, $f_{x,0.01} = 2.1\,a/\pi^{1/2}$, and $f_{x,0.001} = 2.6\,a/\pi^{1/2}$. Figure 2.12 shows the result of using this grid spacing corresponding to $f_{x,0.01}$ as the maximum frequency. Aliasing is not visible in the amplitude plot because the portion of the spectrum that wraps around has a very small value ($0.01\times$ the peak value).

2.5.3 Gaussian signals with quadratic phase

In this case, we add a quadratic phase factor to the Gaussian signal. The Gaussian signal with quadratic phase is defined by

$$g(x) = \exp\left[-\pi\left(ax\right)^2\right]\exp\left[i\pi\left(bx\right)^2\right]. \tag{2.28}$$

This sort of signal arises in the propagation of Gaussian-beam waves. It is mathematically the most general and complicated of the three signals covered in these case studies. Figure 2.13 shows the real and imaginary parts of this signal for the case when $a = 0.25$ and $b = 0.57$. The quadratic phase causes it to oscillate rapidly as $|x|$ increases. The Gaussian amplitude, however, attenuates the oscillations so

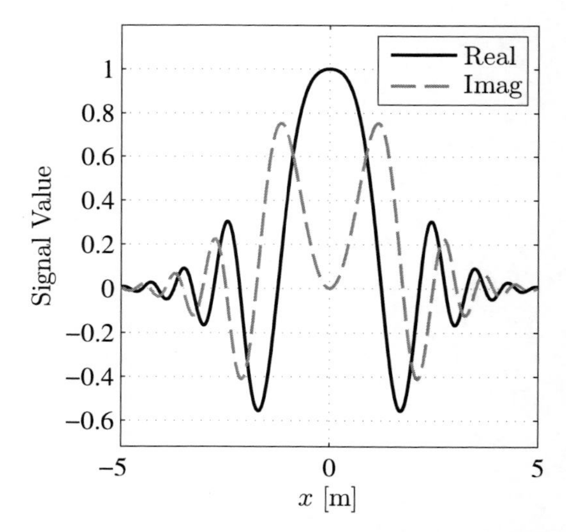

Figure 2.13 Real and imaginary parts of a Gaussian signal with a quadratic phase.

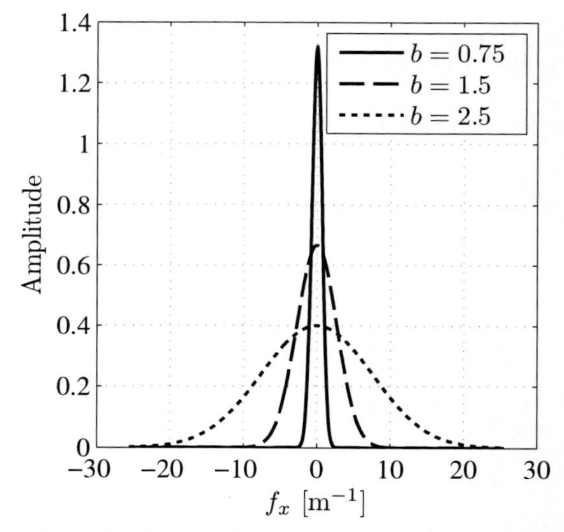

Figure 2.14 Spectral amplitude of a Gaussian signal with a quadratic phase. Clearly, increasing the value of b increases the bandwidth of the signal.

that the signal is in fact nearly bandlimited. To sample this function sufficiently for computing a DFT, we first need to determine the bandwidth of the spectrum. The signal has an analytic FT given by

$$G\left(f_x\right) = \frac{1}{\sqrt{a^2 - ib^2}} \exp\left(-\pi \frac{f_x^2}{a^2 - ib^2}\right). \tag{2.29}$$

Figure 2.14 shows the impact of the curvature parameter b on the width of the spectrum. The plot shows the case of $a = 0.33$ with three different values of b,

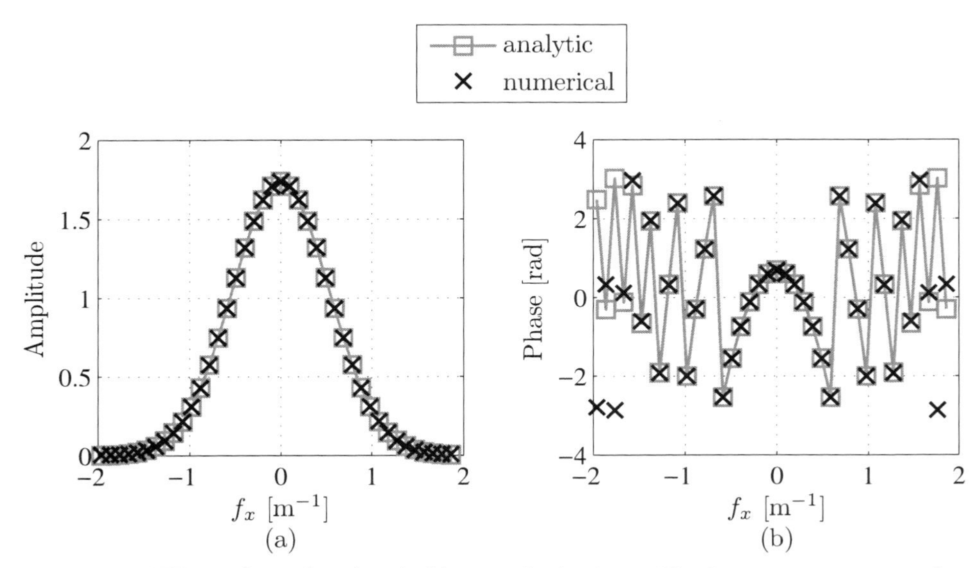

Figure 2.15 DFT of a Gaussian signal with a quadratic phase. The frequency corresponding to $p = 0.01$ was used to compute the grid spacing.

using 0.75, 1.5, and 2.5. The three lines clearly demonstrate that as b increases, so does the width of the spectrum. In fact, we can compute bandwidth from its amplitude using

$$p = \exp\left[-\pi \mathrm{Re}\left(\frac{f_{x,p}^2}{a^2 - ib^2}\right)\right]. \tag{2.30}$$

The result is

$$f_{x,p} = \left[-\left(\frac{a^2 + b^4/a^2}{\pi}\right)\ln p\right]^{1/2}. \tag{2.31}$$

Of course, Eq. (2.31) analytically confirms that $f_{x,p}$ increases with b. Also, note that Eqs. (2.26) and (2.27) are the $b = 0$ cases of Eqs. (2.30) and (2.31).

Figure 2.15 shows the analytic FT and DFT of a Gaussian signal with a quadratic phase. The signal has $a = 0.25$ and $b = 0.57$. It was sampled with grid spacing corresponding to the $p = 0.01$ frequency. Therefore, $f_{x,0.01} = 1.96$ m^{-1} and $\delta = 1/(2f_{x,0.01}) = 0.25$ m, and only 40 samples are required. In the figure, the amplitude clearly matches well, but the DFT phase is slightly inaccurate at the edge of the spatial-frequency grid. If we were simulating a system that could sample no faster than about 1.7 m^{-1}, this would be all right. However, if we needed accuracy at higher spatial frequencies, we might need to do the simulation again with $p = 0.001$.

2.6 Two-Dimensional Discrete Fourier Transforms

We live in a four-dimensional universe (as far as we know) with three spatial dimensions plus time. Optics deals with waves traveling along one spatial dimension, and

Listing 2.5 Code for performing a two-dimensional DFT in MATLAB.

```
1 function G = ft2(g, delta)
2 % function G = ft2(g, delta)
3     G = fftshift(fft2(fftshift(g))) * delta^2;
```

we typically leave off the time dependence. That leaves us working with a function of two spatial dimensions in a plane transverse to the propagation direction. As a result, two-dimensional FTs are used frequently in optics.[8,13] In fact, they are central to the remainder of this book.

To begin studying two-dimensional FTs, we reuse the results of the previous sections with some modifications. We must rewrite Eqs. (2.1) and (2.2), generalizing to two dimensions, as

$$G(f_x, f_y) = \mathcal{F}\{g(x,y)\} = \int_{-\infty}^{\infty}\int_{-\infty}^{\infty} g(x,y)\, e^{-i2\pi(f_x x + f_y y)}\, dx\, dy \tag{2.32}$$

$$g(x,y) = \mathcal{F}^{-1}\{G(f_x, f_y)\} = \int_{-\infty}^{\infty}\int_{-\infty}^{\infty} G(f_x, f_y)\, e^{i2\pi(f_x x + f_y y)}\, df_x\, df_y. \tag{2.33}$$

Then, we make the following changes to Eqs. (2.15)–(2.18):

$$g(x) \Rightarrow g(x,y) \tag{2.34}$$

$$G(f_x) \Rightarrow G(f_x, f_y) \tag{2.35}$$

$$\mathrm{rect}\left(\frac{x}{a}\right) \Rightarrow \mathrm{rect}\left(\frac{x}{a}\right)\mathrm{rect}\left(\frac{y}{b}\right) \tag{2.36}$$

$$a\ \mathrm{sinc}(af_x) \Rightarrow ab\ \mathrm{sinc}(af_x)\,\mathrm{sinc}(bf_y) \tag{2.37}$$

$$a\ \mathrm{comb}(af_x) \Rightarrow ab\ \mathrm{comb}(af_x)\,\mathrm{comb}(bf_y) \tag{2.38}$$

This leads to (assuming same number of grid points, sample size, and spacing in x and y dimensions):

$$\begin{aligned}\widetilde{g}(x,y) = {} & \left[g(x,y)\,\frac{1}{\delta^2}\,\mathrm{comb}\left(\frac{x}{\delta}\right)\mathrm{comb}\left(\frac{y}{\delta}\right)\mathrm{rect}\left(\frac{x}{L}\right)\mathrm{rect}\left(\frac{y}{L}\right)\right] \\ & \otimes \left[\frac{1}{L^2}\,\mathrm{comb}\left(\frac{x}{L}\right)\mathrm{comb}\left(\frac{y}{L}\right)\right]\end{aligned} \tag{2.39}$$

$$\begin{aligned}\widetilde{G}(f_x, f_y) = {} & \left[G(f_x, f_y) \otimes \mathrm{comb}(\delta f_x)\,\mathrm{comb}(\delta f_y) \otimes L^2\ \mathrm{sinc}(Lf_x)\,\mathrm{sinc}(Lf_y)\right] \\ & \times \mathrm{comb}(Lf_x)\,\mathrm{comb}(Lf_y).\end{aligned} \tag{2.40}$$

Listings 2.5–2.6 give MATLAB code for the functions `ft2` and `ift2`, which perform two-dimensional DFTs and DIFTs, respectively. These functions are used frequently throughout the remainder of the book. They are central to two-dimensional convolution, correlation, structure functions, and wave propagation.

Listing 2.6 Code for performing a two-dimensional DIFT in MATLAB.

```
1 function g = ift2(G, delta_f)
2 % function g = ift2(G, delta_f)
3     N = size(G, 1);
4     g = ifftshift(ifft2(ifftshift(G))) * (N * delta_f)^2;
```

2.7 Problems

1. Perform a DFT of $\mathrm{sinc}\,(ax)$ with $a = 1$ and $a = 10$. Plot the results along with the corresponding analytic Fourier transforms.

2. Perform a DFT of $\exp\left(-\pi a^2 x^2\right)$ with $a = 1$ and $a = 10$. Plot the results along with the corresponding analytic Fourier transforms.

3. Perform a DFT of $\exp\left(-\pi a^2 x^2 + i\pi b^2 x^2\right)$ with $a = 1$ and $b = 2$. Plot the results along with the corresponding analytic Fourier transforms.

4. Perform a DFT of $\mathrm{tri}\,(ax)$ with $a = 1$ and $a = 10$. Plot the results along with the corresponding analytic Fourier transforms.

5. Perform a DFT of $\exp\left(-a\,|x|\right)$ with $a = 1$ and $a = 10$. Plot the results along with the corresponding analytic Fourier transforms.

Chapter 3
Simple Computations Using Fourier Transforms

There are many useful computations such as correlations and convolutions that can be implemented using FTs. In fact, taking advantage of computationally efficient DFT techniques such as the FFT often executes much faster than more straightforward implementations. Subsequent chapters reuse these tools in an optical context. For example, convolution is used in Ch. 5 to simulate the effects of diffraction and aberrations on image quality, and structure functions are used in Ch. 9 to validate the statistics of turbulent phase screens.

Three of these tools, namely convolution, correlation, and structure functions, are closely related and have similar mathematical definitions. Furthermore, they are all written in terms of FTs in this chapter. However, their uses are quite different, and each common use is explained in the upcoming sections. These different uses cause the implementations of each to be quite different. For example, correlations and structure functions are usually performed on data that pass through an aperture. Consequently, their computations are modified to remove the effects of the aperture.

The last computation discussed in this chapter is the derivative. Like the other computations in this chapter, the method presented is based on FTs to allow for efficient computation. The method is then generalized to computing gradients of two-dimensional functions. While derivatives and gradients are not used again in later chapters, derivatives are discussed because some readers might want to compute derivatives for topics related to optical turbulence, like simulating the operation of wavefront sensors.

3.1 Convolution

We begin this discussion of FT-based computations with convolution for a couple of reasons. First, convolution plays a central role in linear-systems theory.[14] The output of a linear system is the convolution of the input signal with the system's impulse response. In the context of simulating optical wave propagation, the linear-systems formalism applies to coherent and incoherent imaging, analog optical image processing, and free-space propagation. The second reason we begin with

Listing 3.1 Code for performing a one-dimensional discrete convolution in MATLAB.

```
1 function C = myconv(A, B, delta)
2 % function C = myconv(A, B, delta)
3     N = length(A);
4     C = ift(ft(A, delta) .* ft(B, delta), 1/(N*delta));
```

Listing 3.2 MATLAB example of performing a discrete convolution with comparison to the analytic evaluation of the convolution integral.

```
1 % example_conv_rect_rect.m
2 N = 64;          % number of samples
3 L = 8;           % grid size [m]
4 delta = L / N;   % sample spacing [m]
5 F = 1/L;         % frequency-domain grid spacing [1/m]
6 x = (-N/2 : N/2-1) * delta;
7 w = 2;           % width of rectangle
8 A = rect(x/w);   B = A; % signal
9 C = myconv(A, B, delta); % perform digital convolution
10 % continuous convolution
11 C_cont = w*tri(x/w);
```

convolution is that its practical implementation is the simplest of all the FT-based computations discussed in this chapter.

Throughout this book, we use the symbol $\otimes$ to denote the convolution operation defined by

$$C_{fg}(x) = f(x) \otimes g(x) = \int_{-\infty}^{\infty} f(x')\, g(x - x')\, dx'. \tag{3.1}$$

Often, the two functions being convolved have very different characteristics. Particularly when convolution is used in the context of linear systems, one function is a signal and the other is an impulse response. In the time domain, the impulse response ordinarily has a short duration, while the signal usually has a much longer duration. In the spatial domain, like for optical imaging, the impulse response ordinarily has a narrow spatial extent, while the signal usually occupies a comparatively larger area. The act of convolution smears the input slightly so that the duration or extent of the output is slightly larger than that of the input signal. Often, this spreading effect requires that the inputs to numerical convolution be padded with zeros at the edges of the grid to avoid artifacts of undesired periodicity.[10] In this book, the signals involved are usually already padded with zeros.

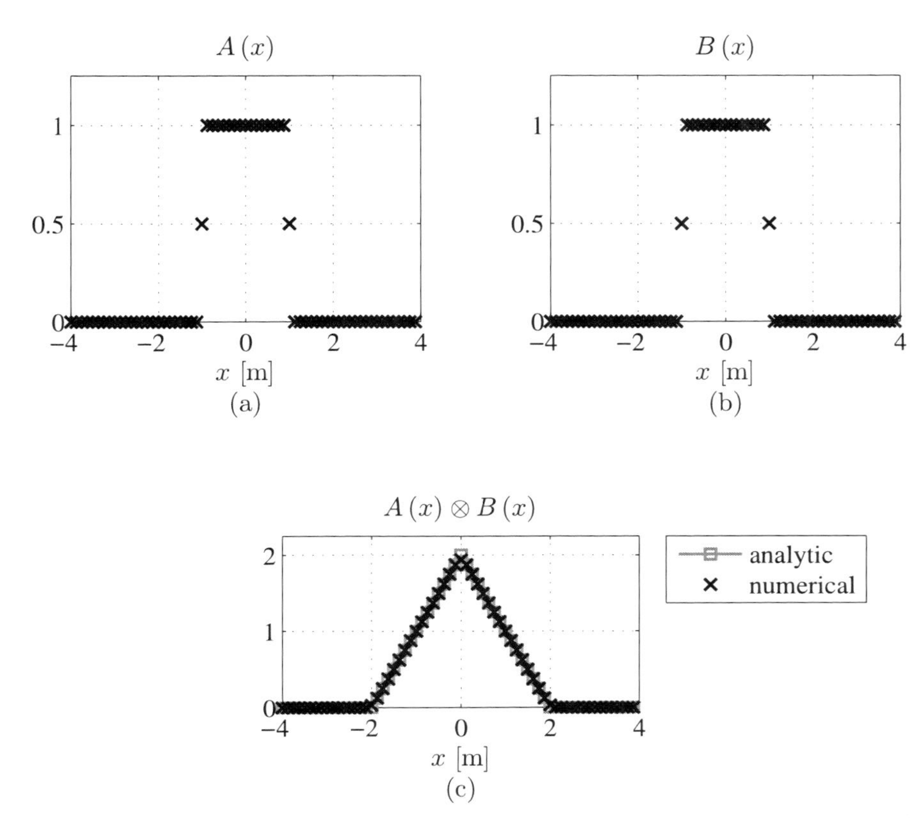

Figure 3.1 A `rect` function convolved with itself. Plots (a) and (b) show the sampled functions that are input into the convolution algorithm. Plot (c) shows the result from the DFT-based computation and the analytic result.

The implementation begins with using the convolution theorem, which is mathematically stated as[5]

$$\mathcal{F}\left[f(x) \otimes g(x)\right] = \mathcal{F}\left[f(x)\right]\mathcal{F}\left[g(x)\right]. \tag{3.2}$$

The beneficial mathematical property here is that the often-difficult-to-compute convolution integral is equivalent to simple multiplication in the frequency domain. Then, by inverse Fourier transforming both sides, Eq. (3.1) can be rewritten as

$$f(x) \otimes g(x) = \mathcal{F}^{-1}\left\{\mathcal{F}\left[f(x)\right]\mathcal{F}\left[g(x)\right]\right\}. \tag{3.3}$$

The computational benefit of the convolution theorem is that, when taking advantage of the FFT algorithm, Eq. (3.3) is typically much faster to evaluate numerically than Eq. (3.1) as a double sum. Accordingly, Listing 3.1 gives MATLAB code for the function `myconv` that takes advantage of this property.

Listing 3.2 gives example use of `myconv`, and the results are plotted in Fig. 3.1. In the example, the function $\mathrm{rect}(x/w)$ is convolved with itself, and the analytic

Listing 3.3 Code for performing a two-dimensional discrete convolution in MATLAB.

```
1 function C = myconv2(A, B, delta)
2 % function C = myconv2(A, B, delta)
3     N = size(A, 1);
4     C = ift2(ft2(A, delta) .* ft2(B, delta), 1/(N*delta));
```

Listing 3.4 MATLAB example of performing a two-dimensional discrete convolution. A rectangle function is convolved with itself.

```
1  % example_conv2_rect_rect.m
2
3  N = 256;       % number of samples
4  L = 16;        % grid size [m]
5  delta = L / N;   % sample spacing [m]
6  F = 1/L;       % frequency-domain grid spacing [1/m]
7  x = (-N/2 : N/2-1) * delta;
8  [x y] = meshgrid(x);
9  w = 2;         % width of rectangle
10 A = rect(x/w) .* rect(y/w);   % signal
11 B = rect(x/w) .* rect(y/w);   % signal
12 C = myconv2(A, B, delta); % perform digital convolution
13 % continuous convolution
14 C_cont = w^2*tri(x/w) .* tri(y/w);
```

result is w tri (x/w). The code uses $w = 2$, a grid size of 8 m, and 64 samples. Clearly, the close agreement between the analytic and numerical results in the figure shows that the computer code is operating properly, and `myconv` uses the proper scaling.

Two-dimensional convolution is quite important in optics. Particularly, to compute a diffraction image, one must convolve the geometric image with the imaging system's two-dimensional spatial impulse response. This optical application of two-dimensional convolution is discussed further in Sec. 5.2. Generalizing Listing 3.1 to perform convolution in two dimensions is quite straightforward. In the computer code, the calls to the functions `ft` and `ift` are replaced by `ft2` and `ift2`, respectively. The MATLAB code is given in Listing 3.3 for the function `myconv2`.

Listing 3.4 gives an example of a two-dimensional convolution. In the example, the function $A(x, y) = \text{rect}(x/w)\,\text{rect}(y/w)$ is convolved with itself. In this case, $w = 2.0$ m, the grid size is 16 m, and there are 256 grid points per side. Figure 3.2 shows the analytic and numerical results. Note the close agreement between them.

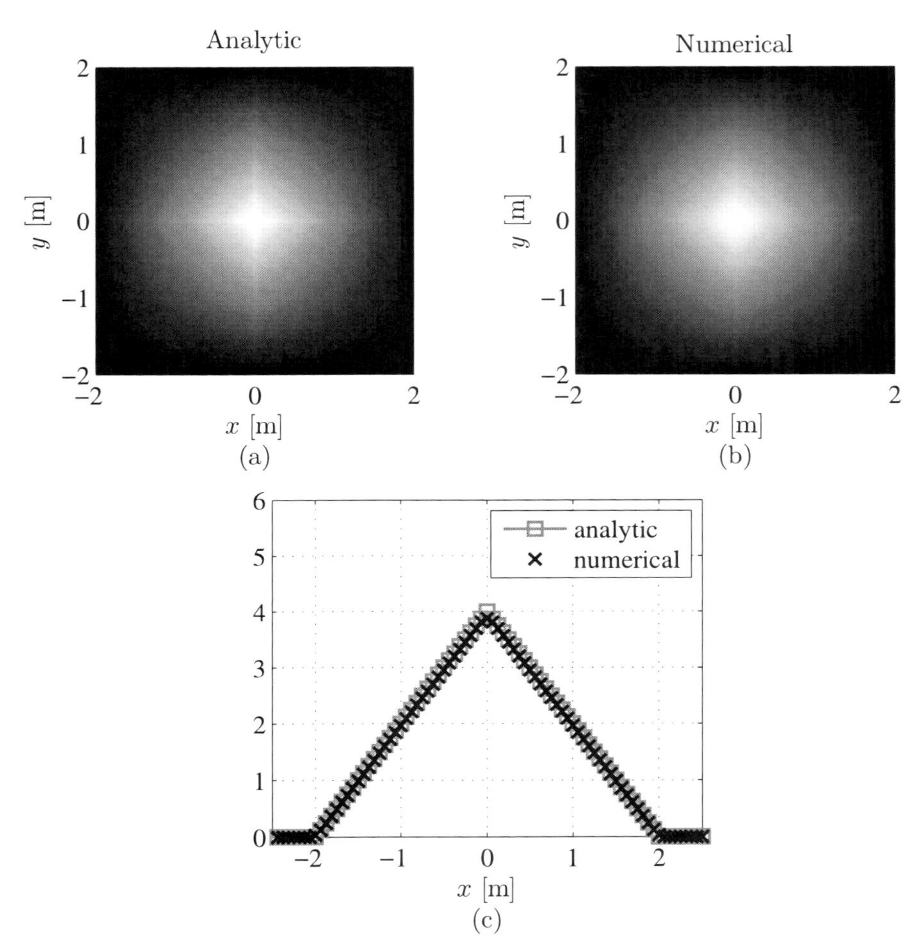

Figure 3.2 A rectangle function convolved with itself. Plot (a) shows the analytic result, while plot (b) shows the numerical result. Plot (c) shows a comparison of the $y = 0$ slices of the analytic and numerical results.

3.2 Correlation

Correlation functions are mathematically very similar to convolutions. Because of the differences in implementation though, we begin the discussion of correlation in two dimensions. Let us define the two-dimensional correlation integral as

$$\Gamma_{fg}(\Delta\mathbf{r}) = f(\mathbf{r}) \star g(\mathbf{r}) = \int_{-\infty}^{\infty} f(\mathbf{r})\, g^*(\mathbf{r} - \Delta\mathbf{r})\, d\mathbf{r}, \tag{3.4}$$

where the $\star$ notation has also been used to denote correlation. Comparing this to Eq. (3.1), we can see that the only mathematical differences between convolution and correlation are a the complex conjugate on $g(x)$ and a minus sign on its argument. There is even a correlation theorem similar to the convolution theorem that provides similar mathematical and computational benefits. Inverse Fourier transforming both sides, Eq. (3.4) can be rewritten as

$$f(x) \star g(x) = \mathcal{F}^{-1}\left\{\mathcal{F}[f(x)]\,\mathcal{F}[g(x)]^*\right\}. \tag{3.5}$$

Despite the mathematical similarities between convolution and correlation, their usages are often quite different. Usually, correlation is often used to determine the similarity between two signals. Accordingly, the two input signals $f(x)$ and $g(x)$ often have relatively similar characteristics, whereas the two inputs to convolution are usually quite different from each other. The separation $\Delta\mathbf{r}$ at which the correlation peaks may tell the distance between features in the two signals. When the two inputs are the same, i.e., $f(x) = g(x)$, it is an auto-correlation. The width of the auto-correlation's peak may reveal information about the signal's variations.

A particular application of auto-correlation in this book is analysis of processes and fields that fluctuate randomly. At any given time or point in space, a random quantity may be specified by a probability density function (PDF). To describe the temporal and spatial variations, often the mean auto-correlation is used. As a relevant example, sometimes optical sources themselves fluctuate randomly. This is the realm of statistical optics.[6] In Ch. 9, the optical field fluctuates randomly due to atmospheric turbulence even if the source field has no fluctuations. The correlation properties of the field contain information about the cause of the fluctuations.[15] For example, Ch. 9 presents theoretical expressions for the mean auto-correlation of optical fields that have propagated through atmospheric turbulence in terms of the turbulent coherence diameter. The theoretical expression is compared to the numerical auto-correlation of simulated random draws of turbulence-degraded fields. The favorable comparison provides a means of verifying proper operation of the turbulent simulation.

The mean correlation is the ensemble average of many independent and identically distributed realizations of Eq. (3.4). The very basic implementation of Eq. (3.4) is very similar to that of convolution. However, optical data are often collected through a circular or annular aperture, while we must represent two-dimensional data in a rectangular array of numbers. Sometimes we wish to isolate the correlation of the data within the pupil and exclude effects of the pupil when we compute quantities like auto-correlation. For example, we may be observing a field that is partially coherent. To relate observation-plane measurements to properties of the source, we need to compute the auto-correlation of the pupil-plane field, not the combined pupil-plane field and aperture. The basic approach, like the one used for convolution, would capture the combined effects of the signal and aperture. To remove the effects of the aperture, the implementation presented here is more complicated than that for convolution.

Let the optical field be $u(\mathbf{r})$ and the shape of the pupil be represented by $w(\mathbf{r})$. The function $w(\mathbf{r})$ is a "window" that is usually equal to one inside the optical aperture and zero outside, written formally as

$$w(\mathbf{r}) = \begin{cases} 1 & \mathbf{r} \text{ inside pupil} \\ 0 & \mathbf{r} \text{ outside pupil.} \end{cases} \quad (3.6)$$

This allows us to use only the region of the field that is transmitted through the aperture. The data that our sensors collect through the aperture is

$$u'(\mathbf{r}) = u(\mathbf{r})\, w(\mathbf{r}). \quad (3.7)$$

If we compute the auto-correlation of the windowed data, we get

$$\Gamma_{u'u'}(\Delta\mathbf{r}) = u'(\mathbf{r}) \star u'(\mathbf{r}) = \int_{-\infty}^{\infty} u(\mathbf{r})\, u^*(\mathbf{r} - \Delta\mathbf{r})\, w(\mathbf{r})\, w^*(\mathbf{r} - \Delta\mathbf{r})\, d\mathbf{r}. \quad (3.8)$$

The integrand is equal to $u(\mathbf{r})\, u^*(\mathbf{r} - \Delta\mathbf{r})$ wherever $w(\mathbf{r})\, w^*(\mathbf{r} - \Delta\mathbf{r})$ is nonzero. Let us denote this region as $R(\mathbf{r}, \Delta\mathbf{r})$. Then we can rewrite the integral as

$$\Gamma_{u'u'}(\Delta\mathbf{r}) = \int_{R(\mathbf{r},\Delta\mathbf{r})} u(\mathbf{r})\, u^*(\mathbf{r} - \Delta\mathbf{r})\, d\mathbf{r}. \quad (3.9)$$

Now we compute the area of $R(\mathbf{r}, \Delta\mathbf{r})$ as

$$A(\Delta\mathbf{r}) = \int R(\mathbf{r}, \Delta\mathbf{r})\, d\mathbf{r} = \Gamma_{ww}(\Delta\mathbf{r}). \quad (3.10)$$

Listing 3.5 MATLAB code for performing a two-dimensional discrete correlation removing effects of the aperture.

```
1  function c = corr2_ft(u1, u2, mask, delta)
2  % function c = corr2_ft(u1, u2, mask, delta)
3
4      N = size(u1, 1);
5      c = zeros(N);
6      delta_f = 1/(N*delta);   % frequency grid spacing [m]
7
8      U1 = ft2(u1 .* mask, delta);     % DFTs of signals
9      U2 = ft2(u2 .* mask, delta);
10     U12corr = ift2(conj(U1) .* U2, delta_f);
11
12     maskcorr = ift2(abs(ft2(mask, delta)).^2, delta_f) ...
13         * delta^2;
14     idx = logical(maskcorr);
15     c(idx) = U12corr(idx) ./ maskcorr(idx) .* mask(idx);
```

Listing 3.6 MATLAB example of performing a two-dimensional discrete auto-correlation. A rectangle function is correlated with itself.

```
1  % example_corr2_rect_rect.m
2
3  N = 256;        % number of samples
4  L = 16;         % grid size [m]
5  delta = L / N;  % sample spacing [m]
6  F = 1/L;        % frequency-domain grid spacing [1/m]
7  x = (-N/2 : N/2-1) * delta;
8  [x y] = meshgrid(x);
9  w = 2;          % width of rectangle
10 A = rect(x/w) .* rect(y/w);   % signal
11 mask = ones(N);
12 % perform digital correlation
13 C = corr2_ft(A, A, mask, delta);
14 % analytic correlation
15 C_cont = w^2*tri(x/w) .* tri(y/w);
```

If we know that the average of $\Gamma_{uu}(\Delta\mathbf{r})$ is truly independent of $\mathbf{r}$, $u(\mathbf{r})$ is called wide-sense stationary, and we can write

$$\langle \Gamma_{u'u'}(\Delta\mathbf{r})\rangle = A(\Delta\mathbf{r})\langle \Gamma_{uu}(\Delta\mathbf{r})\rangle . \tag{3.11}$$

To compute Eq. (3.11) efficiently, we can use FTs. Using the auto-correlation theorem, we can define

$$W(\mathbf{f}) = \mathcal{F}\{w(\mathbf{r})\} \tag{3.12}$$

$$U'(\mathbf{f}) = \mathcal{F}\{u'(\mathbf{r})\}, \tag{3.13}$$

and then write

$$\langle \Gamma_{uu}(\Delta\mathbf{r})\rangle = \frac{\left\langle \mathcal{F}^{-1}\left\{|U'(\mathbf{f})|^2\right\}\right\rangle}{\mathcal{F}^{-1}\left\{|W(\mathbf{f})|^2\right\}} \tag{3.14}$$

Equation (3.14) can be generalized to handle cross correlations between two fields $u_1(\mathbf{r})$ and $u_2(\mathbf{r})$. MATLAB code for computing this cross correlation using a generalized version of Eq. (3.14) is given in Listing 3.5.

Listing 3.6 gives an example of a two-dimensional auto-correlation. In the example, the function $A(x,y) = \mathrm{rect}(x/w)\,\mathrm{rect}(y/w)$ is correlated with itself. In this case, $w = 2.0$ m, the grid size is 16 m, and there are 256 grid points per side. The mask value is one over the entire grid because there is no aperture. Because the function is symmetric about the x and y axes, the result is the same as the convolution example above. Figure 3.3 shows the analytic and numerical results. Once again, note the close agreement between them. An example of computing mean auto-correlation with an aperture mask is given in Sec. 9.5.5.

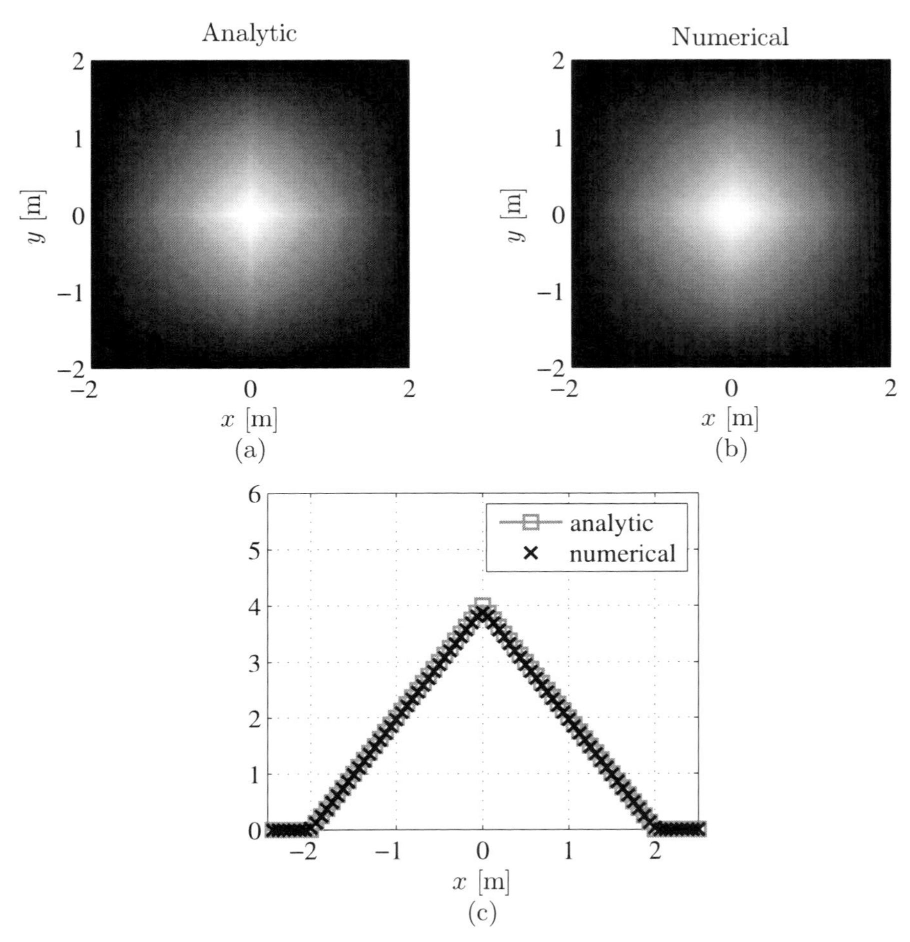

Figure 3.3 A rectangle function correlated with itself. Plot (a) shows the analytic result, while plot (b) shows the numerical result. Plot (c) shows a comparison of the $y = 0$ slices of the analytic and numerical results.

3.3 Structure Functions

Structure functions are another statistical measure of random fields, and they are closely related to auto-correlations. They are particularly appropriate for studying random fields that are not wide-sense stationary. See Ref. 6 for a detailed discussion of statistical stationarity. Structure functions are often used in optical turbulence to describe the behavior of quantities like refractive index, phase, and log-amplitude. The structure function of one realization of a random field $g(\mathbf{r})$ is defined as

$$D_g(\Delta\mathbf{r}) = \int [g(\mathbf{r}) - g(\mathbf{r} + \Delta\mathbf{r})]^2 \, d\mathbf{r}. \tag{3.15}$$

Like with correlations, a statistical structure function is an ensemble average over Eq. (3.15). It can be shown that when the random field is statistically isotropic, the

Listing 3.7 MATLAB code for performing a two-dimensional discrete structure function removing effects of the aperture.

```
1  function D = str_fcn2_ft(ph, mask, delta)
2  % function D = str_fcn2_ft(ph, mask, delta)
3
4      N = size(ph, 1);
5      ph = ph .* mask;
6
7      P = ft2(ph, delta);
8      S = ft2(ph.^2, delta);
9      W = ft2(mask, delta);
10     delta_f = 1/(N*delta);
11     w2 = ift2(W.*conj(W), delta_f);
12
13     D = 2 * ift2(real(S.*conj(W)) - abs(P).^2, ...
14         delta_f) ./ w2 .* mask;
```

Listing 3.8 MATLAB example of performing a two-dimensional structure function of a rectangle function.

```
1  % example_strfcn2_rect.m
2
3  N = 256;        % number of samples
4  L = 16;         % grid size [m]
5  delta = L / N;  % sample spacing [m]
6  F = 1/L;      % frequency-domain grid spacing [1/m]
7  x = (-N/2 : N/2-1) * delta;
8  [x y] = meshgrid(x);
9  w = 2;          % width of rectangle
10 A = rect(x/w) .* rect(y/w);  % signal
11 mask = ones(N);
12 % perform digital structure function
13 C = str_fcn2_ft(A, mask, delta) / delta^2;
14 % continuous structure function
15 C_cont = 2 * w^2 * (1 - tri(x/w) .* tri(y/w));
```

mean structure function and auto-correlation are related by

$$D_g(\Delta\mathbf{r}) = 2\left[\Gamma_{gg}(\mathbf{0}) - \Gamma_{gg}(\Delta\mathbf{r})\right]. \tag{3.16}$$

Also similar to correlations, we often must compute the structure function of windowed data. Using windowed data u' yields

$$\langle D_{u'}(\Delta\mathbf{r})\rangle = A(\Delta\mathbf{r})\langle D_u(\Delta\mathbf{r})\rangle. \tag{3.17}$$

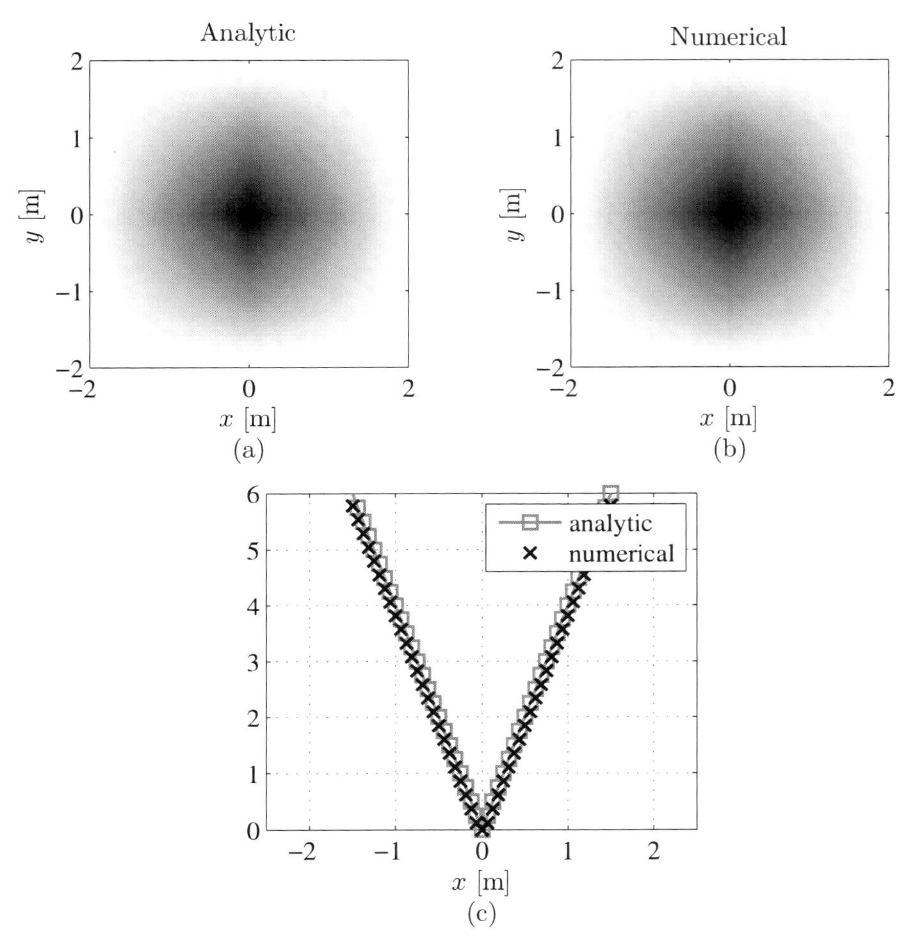

Figure 3.4 Structure function of a rectangle function. Plot (a) shows the analytic result, while plot (b) shows the numerical result. Plot (c) shows a comparison of the $y = 0$ slices of the analytic and numerical results.

Then we must focus on $D_{u'}(\Delta\mathbf{r})$. Multiplying out the terms inside the integral, we get

$$D_{u'}(\Delta\mathbf{r}) = \int \Big[u'^2(\mathbf{r})\, w(\mathbf{r}+\Delta\mathbf{r}) \\ -2u'(\mathbf{r})\, u'(\mathbf{r}+\Delta\mathbf{r}) + u'^2(\mathbf{r}+\Delta\mathbf{r})\, w(\mathbf{r})\Big]\, d\mathbf{r}. \tag{3.18}$$

Now we can replace each term by its Fourier integral representation, which allows for an efficient computation when we use FFTs. To do so, first let us define

$$W(\mathbf{f}) = \mathcal{F}\{w(\mathbf{r})\} \tag{3.19}$$

$$U'(\mathbf{f}) = \mathcal{F}\{u'(\mathbf{r})\} \tag{3.20}$$

$$S(\mathbf{f}) = \mathcal{F}\left\{[u'(\mathbf{r})]^2\right\}. \tag{3.21}$$

Also, note that $W(\mathbf{f}) = W^*(\mathbf{f})$ because $w(\mathbf{r})$ is real. Then, with these definitions and properties, we can write

$$D_{u'}(\Delta\mathbf{r}) = \int \int_{-\infty}^{\infty} \int_{-\infty}^{\infty} \{S(\mathbf{f}_1) W^*(\mathbf{f}_2) + S^*(\mathbf{f}_2) W(\mathbf{f}_1) - 2U'(\mathbf{f}_1) \left[U'(\mathbf{f}_2)\right]^*\} \times e^{i2\pi(\mathbf{f}_1+\mathbf{f}_2)\cdot\mathbf{r}} e^{-i2\pi\mathbf{f}_2\cdot\Delta\mathbf{r}} \, d\mathbf{f}_1 \, d\mathbf{f}_2 \, d\mathbf{r}. \tag{3.22}$$

Now, evaluating the $\mathbf{r}$ integral and then the $\mathbf{f}_2$ integral yields

$$D_{u'}(\Delta\mathbf{r}) = \int_{-\infty}^{\infty} \{S(\mathbf{f}_1) W^*(\mathbf{f}_1) + S^*(\mathbf{f}_1) W(\mathbf{f}_1) - 2U'(\mathbf{f}_1) \left[U'(\mathbf{f}_1)\right]^*\} e^{-i2\pi\mathbf{f}_1\cdot\Delta\mathbf{r}} \, d\mathbf{f}_1 \tag{3.23}$$

$$= 2 \int_{-\infty}^{\infty} \left\{\mathrm{Re}\left[S(\mathbf{f}_1) W^*(\mathbf{f}_1)\right] - \left|U'(\mathbf{f}_1)\right|^2\right\} e^{-i2\pi\mathbf{f}_1\cdot\Delta\mathbf{r}} \, d\mathbf{f}_1 \tag{3.24}$$

$$= 2\mathcal{F}\left\{\mathrm{Re}\left[S(\mathbf{f}_1) W^*(\mathbf{f}_1)\right] - \left|U'(\mathbf{f}_1)\right|^2\right\}. \tag{3.25}$$

Listing 3.7 implements Eqs. (3.17) and (3.25) to compute a structure function through the use of FTs.

Listing 3.8 gives an example of computing a two-dimensional structure function. The example computes the structure function of the two-dimensional signal $A(x, y) = \mathrm{rect}(x/w)\,\mathrm{rect}(y/w)$. As in the previous example, $w = 2.0$ m, the grid size is 16 m, and there are 256 grid points per side. The mask value is one over the entire grid. To compute the analytic result, we can take advantage of the relationship between structure functions and auto-correlations as given by Eq. (3.16). This example uses the same signal as the previous example of correlation, so we apply this relationship to compute the analytic structure function from the analytic auto-correlation. Figure 3.4 shows the analytic and numerical results. Once again, note the close agreement between them. Sections 9.3 and 9.5.5 give examples of computing the mean structure function of a random field.

3.4 Derivatives

This chapter concludes with one last computation based on DFTs, namely derivatives. Derivatives are not used again in this book, but readers who simulate the operation of devices such as wavefront sensors may find this section useful. Several useful devices such as the Shack-Hartmann and shearing-interferometer wavefront sensors can measure the gradient of optical phase.

Listing 3.9 MATLAB code for performing a one-dimensional discrete derivative.

```
function der = derivative_ft(g, delta, n)
% function der = derivative_ft(g, delta, n)

    N = length(g);    % number of samples in g
    % grid spacing in the frequency domain
    F = 1/(N*delta);
    f_X = (-N/2 : N/2-1) * F;    % frequency values

    der = ift((i*2*pi*f_X).^n .* ft(g, delta), F);
```

Listing 3.10 MATLAB example of performing a one-dimensional discrete derivative. The corresponding plots are shown in Fig. 3.5.

```
% example_derivative_ft.m

N = 64;        % number of samples
L = 6;         % grid size [m]
delta = L/N;     % grid spacing [m]
x = (-N/2 : N/2-1) * delta;
w = 3;         % size of window (or region of interest) [m]
window = rect(x/w); % window function [m]
g = x.^5 .* window; % function
% discrete derivatives
gp_samp = real(derivative_ft(g, delta, 1)) .* window;
gpp_samp = real(derivative_ft(g, delta, 2)) .* window;
% analytic derivatives
gp = 5*x.^4 .* window;
gpp = 20*x.^3 .* window;
```

By taking the nth-order derivative of Eq. (2.1) with respect to x and moving the derivative operator inside the FT, it is easy to show that

$$\mathcal{F}\left\{\frac{d^n}{dx^n}g(x)\right\} = (i2\pi f_x)^n\,\mathcal{F}\{g(x)\}\,. \tag{3.26}$$

We can take advantage of this relationship to compute $dg(x)/dx$ by taking the inverse FT of both sides. This is the principle behind the MATLAB code shown in Listing 3.9, which gives the `derivative_ft` function.

Listing 3.10 shows example usage of the `derivative_ft` function. In this example, $g(x) = x^5$. The first two derivatives of this function are computed, and the results are shown in Fig. 3.5 along with the analytic results for comparison.

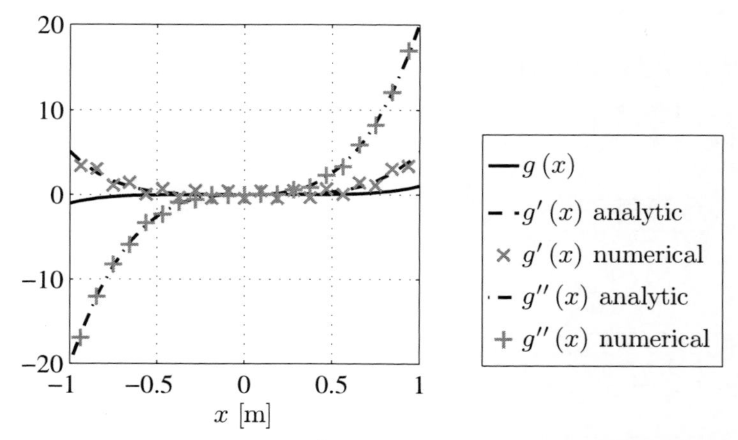

Figure 3.5 Plot of the function $g(x) = x^5$ and its first two derivatives computed numerically with the analytic expressions included for comparison.

Listing 3.11 MATLAB code for computing the discrete gradient of a function using FTs.

```
1  function [gx gy] = gradient_ft(g, delta)
2  % function [gx gy] = gradient_ft(g, delta)
3
4      N =size(g, 1);     % number of samples per side in g
5      % grid spacing in the frequency domain
6      F = 1/(N*delta);
7      fX = (-N/2 : N/2-1) * F;     % frequency values
8      [fX fY] = meshgrid(fX);
9      gx = ift2(i*2*pi*fX .* ft2(g, delta), F);
10     gy = ift2(i*2*pi*fY .* ft2(g, delta), F);
```

Note that a window function is used to limit the extent of the signal and mitigate aliasing of the computed spectrum because $g(x)$ and its first few derivatives are not bandlimited functions. Using the window function improves the accuracy of the numerical derivative.

Now, generalizing Eq. (3.26) to two dimensions, we can compute the x and y partial derivatives of a two-dimensional scalar function $g(x, y)$. Using steps similar to those that produced Eq. (3.26), it is easy to show that

$$\mathcal{F}\left\{\frac{\partial^n}{\partial x^n} g(x, y)\right\} = (i2\pi f_x)^n \mathcal{F}\{g(x, y)\} \tag{3.27}$$

$$\mathcal{F}\left\{\frac{\partial^n}{\partial y^n} g(x, y)\right\} = (i2\pi f_y)^n \mathcal{F}\{g(x, y)\}. \tag{3.28}$$

Then the gradient of the function uses the $n = 1$ case so that

$$\nabla g(x, y) = \mathcal{F}^{-1}\{i2\pi f_x \mathcal{F}\{g(x, y)\}\}\,\hat{\mathbf{i}} + \mathcal{F}^{-1}\{i2\pi f_y \mathcal{F}\{g(x, y)\}\}\,\hat{\mathbf{j}}. \tag{3.29}$$

Listing 3.12 MATLAB example of performing a discrete gradient of a two-dimensional scalar function. The corresponding plots are shown in Fig. 3.6.

```
1  % example_gradient_ft.m
2  N = 64;        % number of samples
3  L = 6;         % grid size [m]
4  delta = L/N;       % grid spacing [m]
5  x = (-N/2 : N/2-1) * delta;
6  [x y] = meshgrid(x);
7  g = exp(-(x.^2 + y.^2));
8  % computed derivatives
9  [gx_samp gy_samp] = gradient_ft(g, delta);
10 gx_samp = real(gx_samp);
11 gy_samp = real(gy_samp);
12 % analytic derivatives
13 gx = -2*x.*exp(-(x.^2+y.^2));
14 gy = -2*y.*exp(-(x.^2+y.^2));
```

This is easily implemented in MATLAB code, as shown in Listing 3.11, which gives the `gradient_ft` function.

Listing 3.12 shows example usage of the `gradient_ft` function. In this example,

$$g(x,y) = \exp\left[-\left(x^2+y^2\right)\right], \tag{3.30}$$

and the analytic gradient is given by

$$\nabla g(x,y) = -2\exp\left[-\left(x^2+y^2\right)\right]\left(x\hat{\mathbf{i}}+y\hat{\mathbf{j}}\right). \tag{3.31}$$

The numerical gradient of this function is computed in the listing, and the results are shown in Fig. 3.6 along with the analytic results for comparison. This time, a window function is not needed because $g(x,y)$ is nearly bandlimited. The quiver plots shown in Figs. 3.6(b) and (c) show the same trends. While it is not exactly evident in the plots, the analytic and numerical gradients are in very close agreement.

3.5 Problems

1. Perform a discrete convolution of the signal function $\mathrm{rect}(x+a)+\mathrm{tri}(x)$ with the impulse response $\exp\left[-(\pi/3)x^2\right]$ for several values of a. At which value of a are the two features in the signal just barely resolved? You do not need to use a formal criterion for resolution, just visually inspect plots of the convolution results.

2. Perform a discrete convolution of the signal $\mathrm{circ}\left[a\left(x^2+y^2\right)^{1/2}\right]$ with itself for $a=1$ and $a=10$. Show the two-dimensional surface plot of the numer-

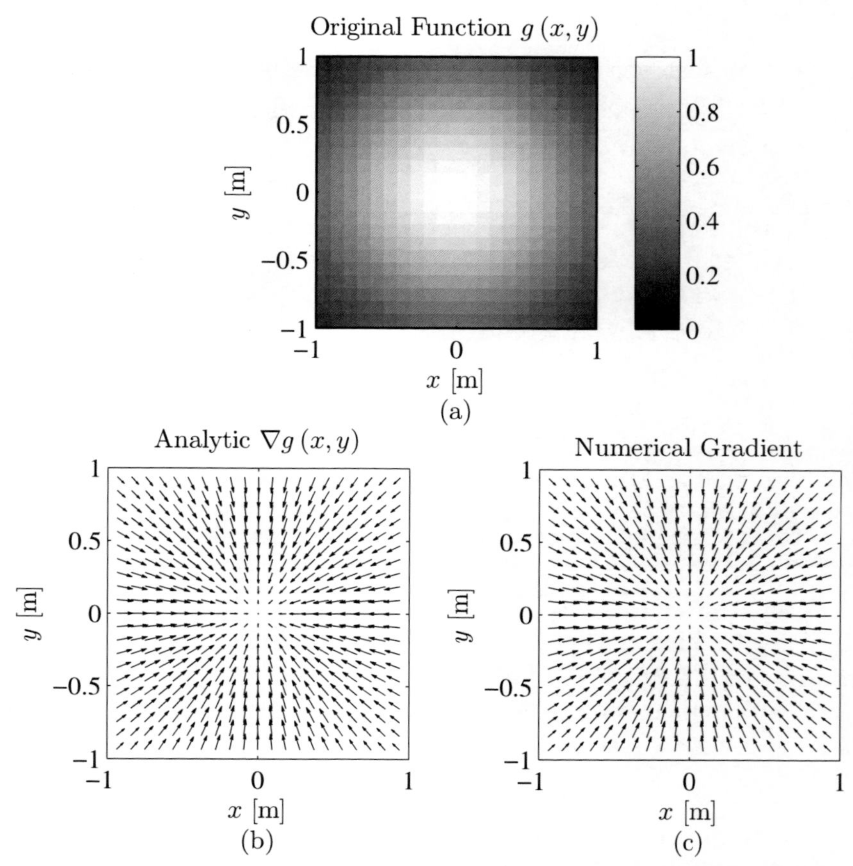

Figure 3.6 Plot of the function $g(x,y) = \exp\left[-\left(x^2+y^2\right)\right]$ and its gradient computed numerically with the analytic expressions included for comparison.

ical and analytic results and a plot of the $y = 0$ slice of the numerical and analytic results.

3. Numerically compute the first derivative of the function $g(x) = J_2(x)$, where $J_2(x)$ is a Bessel function of the first kind, order 2. Plot the result and show agreement with the analytic answer in the region $-1 \leq x \leq 1$.

Chapter 4
Fraunhofer Diffraction and Lenses

To obtain accurate results, evaluating the Fresnel diffraction integral numerically requires some care. Therefore, this chapter first deals with two simpler topics: diffraction with the Fraunhofer approximation and diffraction with lenses. This allows some optical examples of FTs to be demonstrated without the significant algorithm development and sampling analysis required for simulating Fresnel diffraction. Vacuum propagation algorithms and sampling analysis for Fresnel propagation are the subjects Chs. 6–8. Computing diffracted fields in the Fraunhofer approximation or when a lens is present does not require quite so much analysis up front. Additionally, these simple computations involve only a single DFT for each pattern. Chapter 2 provides the requisite background. Consequently, readers may notice that the MATLAB code listings in this chapter are quite simple.

4.1 Fraunhofer Diffraction

When light propagates very far from its source aperture, the optical field in the observation plane is very closely approximated by the Fraunhofer diffraction integral, given in Ch. 1 and repeated here for convenience:

$$U(x_2, y_2) = \frac{e^{ik\Delta z} e^{i\frac{k}{2\Delta z}\left(x_2^2+y_2^2\right)}}{i\lambda\Delta z} \int_{-\infty}^{\infty}\int_{-\infty}^{\infty} U(x_1, y_1)\, e^{-i\frac{k}{\Delta z}(x_1 x_2 + y_1 y_2)}\, dx_1\, dy_1 . \tag{4.1}$$

According to Goodman,[5] "very far" is defined by the inequality

$$\Delta z > \frac{2D^2}{\lambda}, \tag{4.2}$$

where Δz is the propagation distance, D is the diameter of the source aperture, and λ is the optical wavelength. This is a good approximation because the quadratic phase is nearly flat over the source.

The Fraunhofer integral can be cast in the form of an FT that makes use of the

Table 4.1 Definition of symbols for optical propagation.

symbol	meaning
$\mathbf{r}_1 = (x_1, y_1)$	source-plane coordinates
$\mathbf{r}_2 = (x_2, y_2)$	observation-plane coordinates
δ_1	grid spacing in source plane
δ_2	grid spacing in observation plane
Δz	distance between source plane and observation plane

Listing 4.1 Code for performing a Fraunhofer propagation in MATLAB.

```
1  function [Uout x2 y2] = ...
2      fraunhofer_prop(Uin, wvl, d1, Dz)
3  % function [Uout x2 y2] = ...
4  %     fraunhofer_prop(Uin, wvl, d1, Dz)
5
6      N = size(Uin, 1);    % assume square grid
7      k = 2*pi / wvl;   % optical wavevector
8      fX = (-N/2 : N/2-1) / (N*d1);
9      % observation-plane coordinates
10     [x2 y2] = meshgrid(wvl * Dz * fX);
11     clear('fX');
12     Uout = exp(i*k/(2*Dz)*(x2.^2+y2.^2)) ...
13         / (i*wvl*Dz) .* ft2(Uin, d1);
```

lessons from Ch. 2:

$$U(x_2, y_2) = \frac{e^{ik\Delta z} e^{i\frac{k}{2\Delta z}\left(x_2^2+y_2^2\right)}}{i\lambda\Delta z} \mathcal{F}\left\{U(x_1, y_1)\right\}\Big|_{f_{x1}=\frac{x_2}{\lambda\Delta z}, f_{y1}=\frac{y_2}{\lambda\Delta z}}. \tag{4.3}$$

To evaluate this on a grid, we must define the grid properties. We call the grid spacings δ_1 and δ_2 in the source and observation planes, respectively. The spatial-frequency variable for the source plane is $\mathbf{f}_1 = (f_{x1}, f_{y1})$, and its grid spacing is δ_{f1}. Now, the reader should notice that these spatial frequencies are directly mapped to the observation plane's spatial coordinates x_2 and y_2. These symbols are summarized in Table 4.1 and depicted in Fig. 1.2.

Now, numerically evaluating a Fraunhofer diffraction integral is a simple matter of performing an FT with the appropriate multipliers and spatial scaling. Listing 4.1 gives the MATLAB function `fraunhofer_prop` that can be used to numerically perform a wave-optics propagation when the Fraunhofer diffraction integral is valid, i.e., when Eq. (4.2) is true. In the Listing, the factor $\exp(ik\Delta z)$ has been ignored because it is just the on-axis phase. Readers should notice that the code takes advantage of the `ft2` function developed in Ch. 2.

Listing 4.2 demonstrates use of the `fraunhofer_prop` function. The example simulates propagation of a monochromatic plane wave from a circular aperture

Listing 4.2 MATLAB example of simulating a Fraunhofer diffraction pattern with comparison to the analytic result.

```
1  % example_fraunhofer_circ.m
2
3  N = 512;      % number of grid points per side
4  L = 7.5e-3;     % total size of the grid [m]
5  d1 = L / N;   % source-plane grid spacing [m]
6  D = 1e-3;     % diameter of the aperture [m]
7  wvl = 1e-6;   % optical wavelength [m]
8  k = 2*pi / wvl;
9  Dz = 20;        % propagation distance [m]
10
11 [x1 y1] = meshgrid((-N/2 : N/2-1) * d1);
12 Uin = circ(x1, y1, D);
13 [Uout x2 y2] = fraunhofer_prop(Uin, wvl, d1, Dz);
14
15 % analytic result
16 Uout_th = exp(i*k/(2*Dz)*(x2.^2+y2.^2)) ...
17     / (i*wvl*Dz) * D^2*pi/4 ...
18     .* jinc(D*sqrt(x2.^2+y2.^2)/(wvl*Dz));
```

to a distant observation plane. The $y_2 = 0$ slice of the resulting field's amplitude is shown in Fig. 4.1. The numerical results shown in Fig. 4.1 closely match the analytic results. However, if a large region was shown, the edges would begin to show some discrepancy. This is due to aliasing, as discussed in Sec. 2.3. If the example code was modeling a real system with a target board sensor that was only 0.4 m in diameter, then aliasing would not significantly affect the comparison between the numerical prediction and the experimentally measured diffraction pattern. The chosen grid spacing and number of grid points would be sufficient for that purpose.

To state this more concretely, the geometry of the propagation imposes a limit on the observable spatial frequency content of the source. The observation-plane coordinates are related to the spatial frequency of the source via

$$x_2 = \lambda \Delta z f_{x1} \tag{4.4a}$$

$$y_2 = \lambda \Delta z f_{y1}. \tag{4.4b}$$

Then, if a sensor in the $x_2 - y_2$ plane is 0.4 m wide, the maximum values of the observation-plane coordinates are $x_{max} = 0.2$ m and $y_{max} = 0.2$ m. This leads to maximum observable values of the source's spatial frequency $f_{x1,max}$ and $f_{y1,max}$ given by

$$f_{x1,max} = \frac{x_{2,max}}{\lambda \Delta z} \tag{4.5a}$$

$$f_{y1,max} = \frac{y_{2,max}}{\lambda \Delta z}. \tag{4.5b}$$

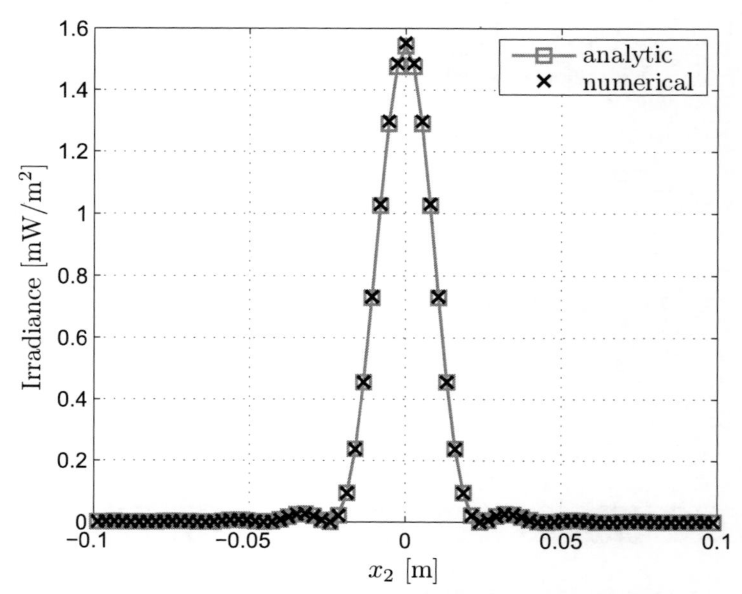

Figure 4.1 The $y_2 = 0$ slice of the amplitude of the Fraunhofer diffraction pattern for a circular aperture. Both the numerical and analytic results are shown for comparison.

As a result, in simulation, propagating a bandlimited (or filtered) version of the real source with spatial frequencies $\leq f_{x1,max}$ and $f_{y1,max}$ would produce the same observation-plane diffraction pattern as one would observe in a laboratory. This principle is used extensively in Ch. 7.

4.2 Fourier-Transforming Properties of Lenses

In this section, the discussion moves to near-field diffraction, governed by the Fresnel diffraction integral in the paraxial approximation for monochromatic waves. This is given in Eq. (1.57) and repeated here for reference:

$$U(x_2, y_2) = \frac{e^{ik\Delta z}}{i\lambda\Delta z} e^{i\frac{k}{2\Delta z}\left(x_2^2+y_2^2\right)} \int\limits_{-\infty}^{\infty}\int\limits_{-\infty}^{\infty} U(x_1, y_1) \times e^{i\frac{k}{2\Delta z}\left(x_1^2+y_1^2\right)} e^{-i\frac{2\pi}{\lambda\Delta z}\left(x_2 x_1+y_2 y_1\right)} \, dx_1 \, dy_1. \tag{4.6}$$

Applying the Fraunhofer approximation in Eq. (4.2) removes the quadratic phase exponential in Eq. (4.6), resulting in the Fraunhofer diffraction integral. However, this approximation is not valid for the scenarios discussed in this section.

In the paraxial approximation, the phase delay imparted by a perfect, spherical (in the paraxial sense), thin lens is given by[5]

$$\phi(x, y) = -\frac{k}{2f_l}\left(x^2 + y^2\right), \tag{4.7}$$

where x and y are coordinates in the exit-pupil plane of the lens, and f_l is the focal length. In this section, a planar transparent object is placed in one of three locations: against (before), the lens, in front of the lens, and behind the lens. The object is illuminated by a normally incident, infinite-extent, uniform-amplitude plane wave. Equation (4.6) is used to propagate the light that passes through the object to the back focal plane of the lens. As a result, the phase term in Eq. (4.7) becomes a part of $U(x_1, y_1)$ inside the Fresnel diffraction integral, resulting in some simplifications as discussed in the next few subsections.

4.2.1 Object against the lens

When the object is placed against the lens as shown in Fig. 4.2, the optical field in the plane just after the lens is

$$U(x_1, y_1) = t_A(x_1, y_1)\, P(x_1, y_1)\, e^{-i\frac{k}{2f_l}\left(x_1^2+y_1^2\right)}, \tag{4.8}$$

where $t_A(x_1, y_1)$ is the aperture transmittance of the object and $P(x_1, y_1)$ is a real function that accounts for apodization by the lens. When Eq. (4.8) is substituted into Eq. (4.6), assuming propagation to the back focal plane, the result is

$$\begin{aligned} U(x_2, y_2) = {} & \frac{1}{i\lambda f_l} e^{i\frac{k}{2f_l}\left(x_2^2+y_2^2\right)} \int_{-\infty}^{\infty}\int_{-\infty}^{\infty} t_A(x_1, y_1) \\ & \times P(x_1, y_1)\, e^{-i\frac{2\pi}{\lambda f_l}\left(x_2 x_1+y_2 y_1\right)}\, dx_1\, dy_1. \end{aligned} \tag{4.9}$$

Like in Sec. 4.1, this can be cast in terms of an FT so that

$$U(x_2, y_2) = \frac{1}{i\lambda f_l} e^{i\frac{k}{2f_l}\left(x_2^2+y_2^2\right)} \mathcal{F}\left\{t_A(x_1, y_1)\, P(x_1, y_1)\right\}\Big|_{f_x=\frac{x_2}{\lambda f_l},\, f_y=\frac{y_2}{\lambda f_l}}. \tag{4.10}$$

This is not an exact FT relationship because of the quadratic phase factor outside the integral. Nonetheless, we can use a DFT to compute diffracted field.

Listing 4.3 gives the MATLAB function `lens_against_ft` from the object plane to the focal plane for an object placed against a converging lens. Notice that the implementation is very similar to `fraunhofer_prop`, which takes advantage of the function `ft2`.

4.2.2 Object before the lens

A more general situation is obtained when the object is placed a distance d before the lens as shown in Fig. 4.3. When the light propagates to the focal plane, the result is

$$U(x_2, y_2) = \frac{1}{i\lambda f_l} e^{i\frac{k}{2f_l}\left(1-\frac{d}{f_l}\right)\left(x_2^2+y_2^2\right)} \int_{-\infty}^{\infty}\int_{-\infty}^{\infty} t_A(x_1, y_1)$$

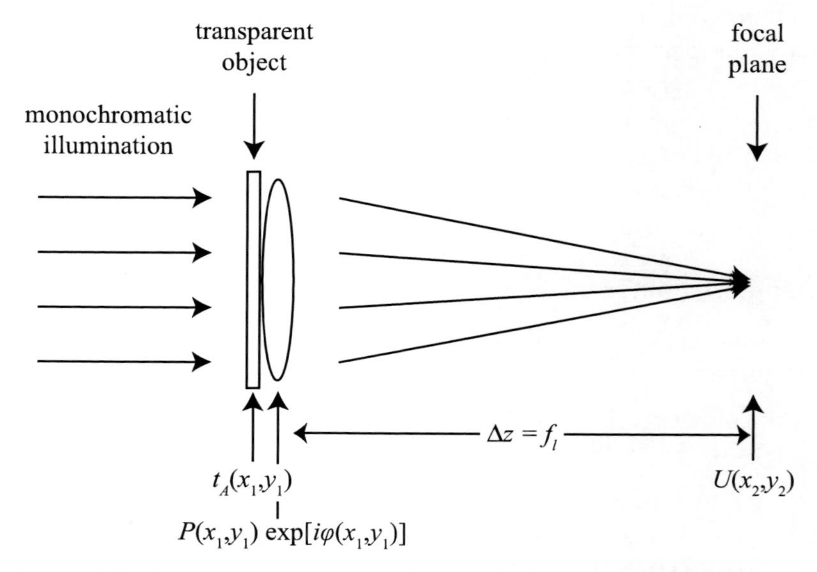

Figure 4.2 Diagram of lens geometry for an object placed against the lens.

Listing 4.3 Code for performing a propagation from the pupil plane to the focal plane for an object placed against (and just before a lens) in MATLAB.

```
1  function [Uout x2 y2] = ...
2      lens_against_ft(Uin, wvl, d1, f)
3  % function [Uout x2 y2] = ...
4  %     lens_against_ft(Uin, wvl, d1, f)
5
6      N = size(Uin, 1);    % assume square grid
7      k = 2*pi/wvl;     % optical wavevector
8      fX = (-N/2 : 1 : N/2 - 1) / (N * d1);
9      % observation plane coordinates
10     [x2 y2] = meshgrid(wvl * f * fX);
11     clear('fX');
12
13     % evaluate the Fresnel-Kirchhoff integral but with
14     % the quadratic phase factor inside cancelled by the
15     % phase of the lens
16     Uout = exp(i*k/(2*f)*(x2.^2 + y2.^2)) ...
17         / (i*wvl*f) .* ft2(Uin, d1);
```

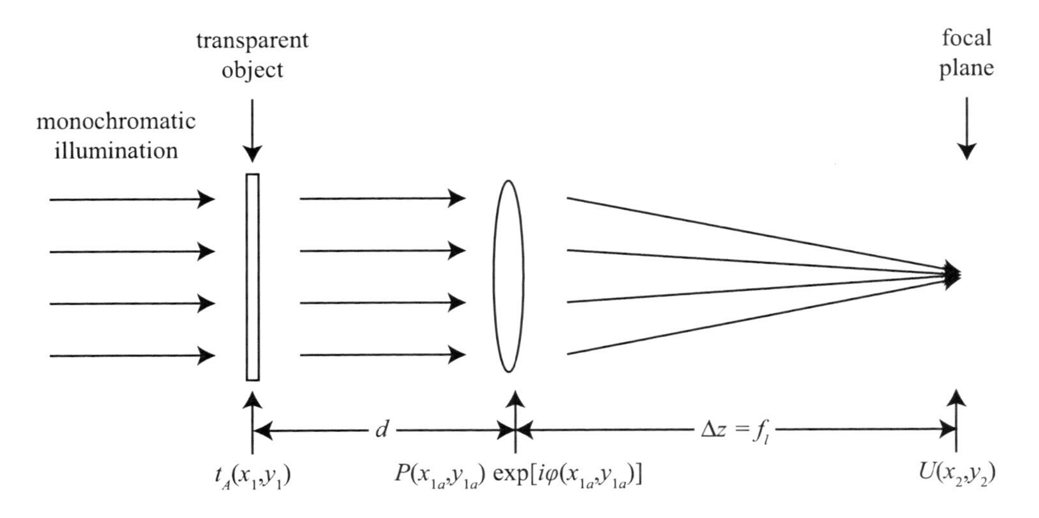

Figure 4.3 Diagram of lens geometry for an object placed before the lens.

$$\times P\left(x_1 + \frac{d}{f_l}x_2, y_1 + \frac{d}{f_l}y_2\right) e^{-i\frac{2\pi}{\lambda f_l}(x_2 x_1 + y_2 y_1)}\, dx_1\, dy_1, \quad (4.11)$$

where the shifted argument of the pupil function accounts for vignetting of the object by the lens aperture. Each point in the focal plane experiences different vignetting with the least occurring for the point on the optical axis. The reader is referred to Goodman (Ref. 5) for more detail. Like in Sec. 4.1, this can be cast in terms of an FT so that

$$\begin{aligned} U(x_2, y_2) = &\frac{1}{i\lambda f_l} e^{i\frac{k}{2f_l}\left(1-\frac{d}{f_l}\right)\left(x_2^2+y_2^2\right)} \\ &\times \mathcal{F}\left\{t_A(x_1, y_1)\, P\left(x_1 + \frac{d}{f_l}x_2, y_1 + \frac{d}{f_l}y_2\right)\right\}\Bigg|_{f_x=\frac{x_2}{\lambda f_l}, f_y=\frac{y_2}{\lambda f_l}}. \end{aligned} \quad (4.12)$$

There two are interesting cases. First, when the object is placed against the lens, $d = 0$, and so Eq. (4.12) reduces to the solution found in Eq. (4.10). Second, when the object is placed in the front focal plane of the lens, $d = f_l$, so the exponential phase factor outside of the integral becomes 1, leaving an exact FT relationship. Listing 4.4 gives the MATLAB function `lens_in_front` for an object placed a distance d in front of a converging lens.

4.2.3 Object behind the lens

When the object is placed behind the lens a distance d away from the focal plane as shown in Fig. 4.4, the optical field $U_s(x_1, y_1)$ just before the object is (in the geometric-optics approximation) a converging spherical wave given by

$$U_s(x_1, y_1) = \frac{f_l}{d} P\left(\frac{f_l}{d}x_1, \frac{f_l}{d}y_1\right) e^{-i\frac{k}{2d}\left(x_1^2+y_1^2\right)}. \quad (4.13)$$

Listing 4.4 Code for performing a propagation from the pupil plane to the focal plane for an object placed in front of a lens in MATLAB.

```
1  function [x2 y2 Uout] ...
2      = lens_in_front_ft(Uin, wvl, d1, f, d)
3  % function [x2 y2 U_out] ...
4  %     = lens_in_front_ft(Uin, wvl, d1, f, d)
5
6      N = size(Uin, 1);    % assume square grid
7      k = 2*pi/wvl;     % optical wavevector
8      fX = (-N/2 : 1 : N/2 - 1) / (N * d1);
9      % observation plane coordinates
10     [x2 y2] = meshgrid(wvl * f * fX);
11     clear('fX');
12
13     % evaluate the Fresnel-Kirchhoff integral but with
14     % the quadratic phase factor inside cancelled by the
15     % phase of the lens
16     Uout = 1 / (i*wvl*f) ...
17         .* exp(i*k/(2*f) * (1-d/f) * (x2.^2 + y2.^2)) ...
18         .* ft2(Uin, d1);
```

This is valid when the distance $d \ll f_l$. Then, the field just after the object is

$$U(x_1, y_1) = \frac{f_l}{d} P\left(\frac{f_l}{d}x_1, \frac{f_l}{d}y_1\right) e^{-i\frac{k}{2d}\left(x_1^2+y_1^2\right)} t_A(x_1, y_1). \tag{4.14}$$

Finally, propagating from the object to the focal plane using Eq. (4.6) yields

$$U(x_2, y_2) = \frac{f_l}{d}\frac{1}{i\lambda d} e^{i\frac{k}{2d}\left(x_2^2+y_2^2\right)} \times \int_{-\infty}^{\infty}\int_{-\infty}^{\infty} t_A(x_1, y_1) P\left(\frac{f_l}{d}x_1, \frac{f_l}{d}y_1\right) e^{-i\frac{2\pi}{\lambda d}\left(x_2 x_1 + y_2 y_1\right)} dx_1\, dy_1. \tag{4.15}$$

As before, this can be cast in terms of an FT so that

$$U(x_2, y_2) = \frac{f_l}{d}\frac{1}{i\lambda d} e^{i\frac{k}{2d}\left(x_2^2+y_2^2\right)} \mathcal{F}\left\{t_A(x_1, y_1) P\left(\frac{f_l}{d}x_1, \frac{f_l}{d}y_1\right)\right\}\Bigg|_{f_x=\frac{x_2}{\lambda d}, f_y=\frac{y_2}{\lambda d}}. \tag{4.16}$$

Listing 4.5 gives the MATLAB function `lens_behind_ft` from the object plane to the focal plane.

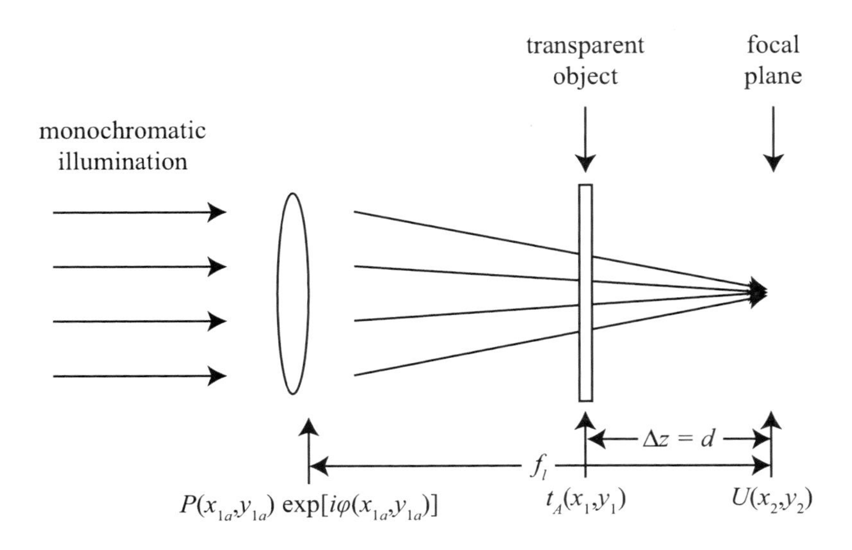

Figure 4.4 Diagram of lens geometry for an object placed behind the lens.

Listing 4.5 Code for performing a propagation from the pupil plane to the focal plane for an object placed behind a converging lens in MATLAB.

```
1  function [x2 y2 Uout] ...
2      = lens_behind_ft(Uin, wvl, d1, f)
3  % function [x2 y2 Uout] ...
4  %     = lens_behind_ft(Uin, wvl, d1, d, f)
5
6      N = size(Uin, 1);    % assume square grid
7      k = 2*pi/wvl;     % optical wavevector
8      fX = (-N/2 : 1 : N/2 - 1) / (N * d1);
9      % observation plane coordinates
10     [x2 y2] = meshgrid(wvl * d * fX);
11     clear('fX');
12
13     % evaluate the Fresnel-Kirchhoff integral but with
14     % the quadratic phase factor inside cancelled by the
15     % phase of the lens
16     Uout = f/d * 1  / (i*wvl*d)...
17         .* exp(i*k/(2*d)*(x2.^2 + y2.^2)) .* ft2(Uin, d1);
```

4.3 Problems

1. Repeat the example in Sec. 4.1 for a 1 mm $\times$ 1 mm square aperture in the source plane. Show the numerical and analytic results together on the same plot.

2. Repeat the example in Sec. 4.1 for a two-slit aperture consisting of two 1 mm $\times$ 1 mm square apertures spaced 0.5 mm apart in the source plane. Show the numerical and analytic results together on the same plot.

3. Repeat the example in Sec. 4.1 for a 1 mm $\times$ 1 mm square amplitude grating in the source plane. Let the amplitude transmittance be

$$t_A(x_1, y_1) = \frac{1}{2}\left[1 + \cos\left(2\pi f_0 x_1\right)\right] \text{rect}\left(\frac{x_1}{D}\right) \text{rect}\left(\frac{y_1}{D}\right), \tag{4.18}$$

where $f_0 = 10/D$. Show the numerical and analytic results together on the same plot.

4. Repeat the example in Sec. 4.1 for a 1 mm $\times$ 1 mm square phase grating in the source plane. Let the amplitude transmittance be

$$t_A(x_1, y_1) = e^{i2\pi\cos(2\pi f_0 x_1)} \text{rect}\left(\frac{x_1}{D}\right) \text{rect}\left(\frac{y_1}{D}\right), \tag{4.19}$$

where $f_0 = 10/D$. Show the numerical and analytic results together on the same plot.

5. A 1-μm wavelength Gaussian laser beam is normally incident on a lens. The beam waist is at the lens with width $w = 2$ cm, and the lens's focal length is 1 m. Assuming that the lens has an infinite diameter, numerically and analytically compute the diffraction pattern in the focal plane. Show the numerical and analytic results together on the same plot.

Chapter 5
Imaging Systems and Aberrations

At the surface, numerically evaluating imaging systems with monochromatic light is a simple extension of two-dimensional discrete convolution, as discussed in Sec. 3.1. This is because the response of light to an imaging system, whether the light is coherent or incoherent, can be modeled as a linear system. Determining the impulse response of an imaging system is more complicated, particularly when the system does not perfectly focus the image. This happens because of aberrations present in the imaging system. In this chapter, aberrations are treated first. Then, we show how aberrations affect the impulse response of imaging systems. Finally, the chapter finishes with a discussion of imaging system performance.

5.1 Aberrations

The light from an extended object can be treated as a continuum of point sources. Each point source emits rays in all directions as shown in Fig. 5.1. In geometric optics, the rays from a given object point that pass all the way through an ideal imaging system are focused to another point. Each point of the object emits (or reflects) an optical field which becomes a diverging spherical wave at the entrance pupil of the imaging system. To focus the light to a point in the image plane, the imaging system must apply a spherical phase delay to convert a diverging spherical

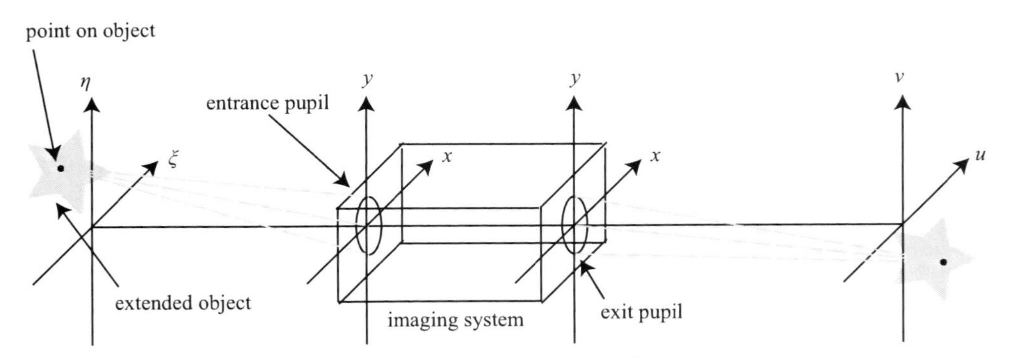

Figure 5.1 Basic model of an imaging system.

Table 5.1 Some Seidel aberration terms and their names.

Term	Name
A_0	piston
$A_1 r\cos\theta + A_2 r\sin\theta$	tilt
$A_3 r^2$	defocus
$A_4 r^2\cos(2\theta) + A_5 r^2\sin(2\theta)$	astigmatism
$A_6 r^3\cos\theta + A_7 r^3\sin\theta$	coma
$A_8 r^4$	spherical aberration

wavefront into a converging spherical wavefront. Aberrations are deviations from the spherical phase delay that cause the rays from a given object point to misfocus and form a finite-sized spot. When the image is viewed as a whole, the aberration manifests itself as blur. Light from different object points can experience different aberrations in the image plane depending on their distance from the optical axis. However, for the purposes of this book, we are not concerned with these field-angle-dependent aberrations but assume that they are constant.

With a detailed description of an imaging system, ray tracing can be used to determine the wavefront aberration for a given object point. Optical design software programs like CODE V,[16] OSLO,[17] and ZEMAX[18] are excellent for this task. In this book, we simply assume that ray tracing has been done already and use the resulting aberration as is. Aberrations can be expressed as a wavefront $W(x,y)$ measured in waves, or optical phase $\phi(x,y) = 2\pi W(x,y)$ measured in radians. Then, we can write a generalized pupil function $\mathcal{P}(x,y)$ by combining the effects of apodization and aberrations into one complex function:

$$\mathcal{P}(x,y) = P(x,y)\, e^{i2\pi W(x,y)}. \tag{5.1}$$

5.1.1 Seidel aberrations

It is common to write an arbitrary wavefront aberration as a polynomial expansion according to

$$\begin{aligned} W(x,y) = {} & A_0 + A_1 r\cos\theta + A_2 r\sin\theta + A_3 r^2 + A_4 r^2\cos(2\theta) \\ & + A_5 r^2\sin(2\theta) + A_6 r^3\cos\theta + A_7 r^3\sin\theta + A_8 r^4 + \ldots \end{aligned} \tag{5.2}$$

where r is a polar normalized pupil coordinate. The normalized coordinate is the physical radial coordinate divided by the pupil radius so that $r = 1$ at the edge of the aperture. These terms are classified as shown in Table 5.1. The A_i coefficients may be field-angle-dependent, but we assume that they are constant when imaging simulations are discussed in Sec. 5.2. If each object point experiences different aberrations, then each image of each object point must be simulated separately.

5.1.2 Zernike circle polynomials

The polynomial expansion from the previous section is convenient because of its simplicity, and it follows directly from use of ray tracing. However, its mathematical properties are lacking. When aberrations become complicated, it is better to use a representation that has completeness and orthogonality, so we describe such a representation here. Most of the time, we deal with circular apertures, and the above polynomial expansion is not orthogonal over a circular aperture. However, Zernike circle polynomials are complete and orthogonal over a circular aperture. Note that there are also Zernike annular polynomials that are orthogonal over an annular aperture, Zernike-Gauss circle polynomials that are orthogonal over a Gaussian aperture, and Zernike-Gauss annular polynomials that are orthogonal over Gaussian, annular apertures.[19] There are even Zernike vector polynomials whose dot product is orthonormal over a circular aperture.[20,21] These are all very interesting and useful, but we discuss only Zernike circle polynomials here.

There are several conventions and ordering schemes for defining Zernike circle polynomials.[4,19,22,23] This book uses the convention of Noll.[22] In this convention, the polynomials are defined as

$$Z_n^m(r,\theta) = \sqrt{2(n+1)}\, R_n^m(r)\, G^m(\theta), \tag{5.3}$$

where m and n are non-negative integers, and $m \leq n$. However, it is convenient to write $Z_n^m(r,\theta)$ with just one index

$$Z_i(r,\theta) = \begin{cases} \sqrt{2(n+1)}\, R_n^m(r)\, G^m(\theta) & m \neq 0 \\ R_n^0(r) & m = 0 \end{cases}. \tag{5.4}$$

The mapping of $(n,m) \rightarrow i$ is complicated, but the ordering for the first 36 Zernike polynomials is given in Table 5.2. The radial and azimuthal factors $R_n^m(r)$ and $G^m(\theta)$ are given by[23]

$$R_n^m(r) = \sum_{s=0}^{(n-m)/2} \frac{(-1)^s (n-s)!}{s!\left(\frac{n+m}{2}-s\right)!\left(\frac{n-m}{2}-s\right)!} r^{n-2s} \tag{5.5a}$$

$$G^m(\theta) = \begin{cases} \sin(m\theta) & i \text{ odd} \\ \cos(m\theta) & i \text{ even.} \end{cases} \tag{5.5b}$$

Listing 5.1 gives the MATLAB function `zernike` that evaluates Eq. (5.4) given the mode number i and normalized polar coordinates on the unit circle. The reader should note that the factorials in Eq. (5.5) are coded in MATLAB as gamma functions $[s! = \Gamma(s+1)]$ because the `gamma` function executes much faster than the `factorial` function.

Figure 5.2 shows an example of three different Zernike polynomials. The particular aberrations shown are three different orders of x primary astigmatism. In plot (a), $n = 2$ and $m = 2$; in plot (b), $n = 4$ and $m = 2$; and in plot (c), $n = 6$ and $m = 2$. Consequently, all three plots have the same azimuthal dependence, $\cos(2\theta)$, while the radial dependence is different for each. The largest power on

Table 5.2 The first 36 Zernike polynomials

n	m	i	$Z_n^m(r,\theta)$	Name
0	0	1	1	piston
1	1	2	$2\,r\cos\theta$	x tilt
1	1	3	$2\,r\sin\theta$	y tilt
2	0	4	$\sqrt{3}\,(2r^2-1)$	defocus
2	2	5	$\sqrt{6}\,r^2\sin(2\theta)$	y primary astigmatism
2	2	6	$\sqrt{6}\,r^2\cos(2\theta)$	x primary astigmatism
3	1	7	$\sqrt{8}\,(3r^3-2r)\sin\theta$	y primary coma
3	1	8	$\sqrt{8}\,(3r^3-2r)\cos\theta$	x primary coma
3	3	9	$\sqrt{8}\,r^3\sin(3\theta)$	y trefoil
3	3	10	$\sqrt{8}\,r^3\cos(3\theta)$	x trefoil
4	0	11	$\sqrt{5}\,(6r^4-6r^2+1)$	primary spherical
4	2	12	$\sqrt{10}\,(4r^4-3r^2)\cos(2\theta)$	x secondary astigmatism
4	2	13	$\sqrt{10}\,(4r^4-3r^2)\sin(2\theta)$	y secondary astigmatism
4	4	14	$\sqrt{10}\,r^4\cos(4\theta)$	x tetrafoil
4	4	15	$\sqrt{10}\,r^4\sin(4\theta)$	y tetrafoil
5	1	16	$\sqrt{12}\,(10r^5-12r^3+3r)\cos\theta$	x secondary coma
5	1	17	$\sqrt{12}\,(10r^5-12r^3+3r)\sin\theta$	y secondary coma
5	3	18	$\sqrt{12}\,(5r^5-4r^3)\cos(3\theta)$	x secondary trefoil
5	3	19	$\sqrt{12}\,(5r^5-4r^3)\sin(3\theta)$	y secondary trefoil
5	5	20	$\sqrt{12}\,r^5\cos(5\theta)$	x pentafoil
5	5	21	$\sqrt{12}\,r^5\sin(5\theta)$	y pentafoil
6	0	22	$\sqrt{7}\,(20r^6-30r^4+12r^2-1)$	secondary spherical
6	2	23	$\sqrt{14}\,(15r^6-20r^4+6r^2)\sin(2\theta)$	y tertiary astigmatism
6	2	24	$\sqrt{14}\,(15r^6-20r^4+6r^2)\cos(2\theta)$	x tertiary astigmatism
6	4	25	$\sqrt{14}\,(6r^6-5r^4)\sin(4\theta)$	y secondary tetrafoil
6	4	26	$\sqrt{14}\,(6r^6-5r^4)\cos(4\theta)$	x secondary tetrafoil
6	6	27	$\sqrt{14}\,r^6\sin(6\theta)$	
6	6	28	$\sqrt{14}\,r^6\cos(6\theta)$	
7	1	29	$4\,(35r^7-60r^5+30r^3-4r)\sin\theta$	y tertiary coma
7	1	30	$4\,(35r^7-60r^5+30r^3-4r)\cos\theta$	x tertiary coma
7	3	31	$4\,(21r^7-30r^5+10r^3)\sin(3\theta)$	
7	3	32	$4\,(21r^7-30r^5+10r^3)\cos(3\theta)$	
7	5	33	$4\,(7r^7-6r^5)\sin(5\theta)$	
7	5	34	$4\,(7r^7-6r^5)\cos(5\theta)$	
7	7	35	$4\,r^7\sin(7\theta)$	
7	7	36	$4\,r^7\cos(7\theta)$	
8	0	37	$3\,(70r^8-140r^6+90r^4-20r^2+1)$	tertiary spherical

Listing 5.1 Code for evaluating Zernike polynomials in MATLAB.

```
1  function Z = zernike(i, r, theta)
2  % function Z = zernike(i, r, theta)
3  % Creates the Zernike polynomial with mode index i,
4  % where i = 1 corresponds to piston
5  load('zernike_index'); % load the mapping of (n,m) to i
6  n = zernike_index(i,1);
7  m = zernike_index(i,2);
8  if m==0
9      Z = sqrt(n+1)*zrf(n,0,r);
10 else
11     if mod(i,2) == 0 % i is even
12         Z = sqrt(2*(n+1))*zrf(n,m,r) .* cos(m*theta);
13     else % i is odd
14         Z = sqrt(2*(n+1))*zrf(n,m,r) .* sin(m*theta);
15     end
16 end
17 return
18
19 % Zernike radial function
20 function R = zrf(n, m, r)
21 R = 0;
22 for s = 0 : ((n-m)/2)
23     num = (-1)^s * gamma(n-s+1);
24     denom = gamma(s+1) * gamma((n+m)/2-s+1) ...
25         * gamma((n-m)/2-s+1);
26     R = R + num / denom * r.^(n-2*s);
27 end
```

each is 2, 4, and 6 for primary, secondary, and tertiary astigmatism, respectively. As we follow the radial portion of each mode from the center to edge of the pupil, the higher-order modes have more peaks, troughs, and zero crossings.

With the modes completely defined, any wavefront $W(r,\theta)$ can be written as a Zernike series with coefficients a_i given by

$$W(r,\theta) = \sum_{i=1}^{\infty} a_i Z_i(r,\theta). \tag{5.6}$$

There are many benefits of this representation, and they are discussed below.

The key property of Zernike polynomials is that they are orthogonal over the unit circle. The orthogonality relationship for this convention of Zernike polynomials is

$$\int_0^1 R_n^m(r)\, R_{n'}^m(r)\, r\, dr = \frac{1}{2n+1}\delta_{nn'} \tag{5.7}$$

$$\int_0^{2\pi} G^m(\theta)\, G^{m'}(\theta)\, d\theta = \pi\delta_{mm'} \tag{5.8}$$

$$\Rightarrow \int_0^{2\pi}\int_0^1 Z_i(r,\theta)\, Z_{i'}(r,\theta)\, r\, dr\, d\theta = \pi\delta_{nn'}\delta_{mm'} = \pi\delta_{ii'}. \tag{5.9}$$

Using the orthogonality relationship, a given wavefront can be decomposed into its Zernike series by computing its Zernike coefficients with

$$a_i = \frac{\int_0^{2\pi}\int_0^1 W(r,\theta)\, Z_i(r,\theta)\, r\, dr\, d\theta}{\int_0^{2\pi}\int_0^1 Z_i^2(r,\theta)\, r\, dr\, d\theta}. \tag{5.10}$$

Often, we have a representation of a two-dimensional wavefront on a sampled two-dimensional Cartesian grid, either from a simulation or measurement. In that case, we can rewrite Eq. (5.10) as a discrete sum over Cartesian coordinates x_p and y_q given by

$$a_i = \frac{\sum_p\sum_q W(x_p,y_q)\, Z_i(x_p,y_q)}{\sum_p\sum_q Z_i^2(x_p,y_q)}. \tag{5.11}$$

In Eq. (5.11) the sums run over all p and q that are within the optical aperture. Notice that Eq. (5.11) does not actually depend on the values of x_p and y_q, only the

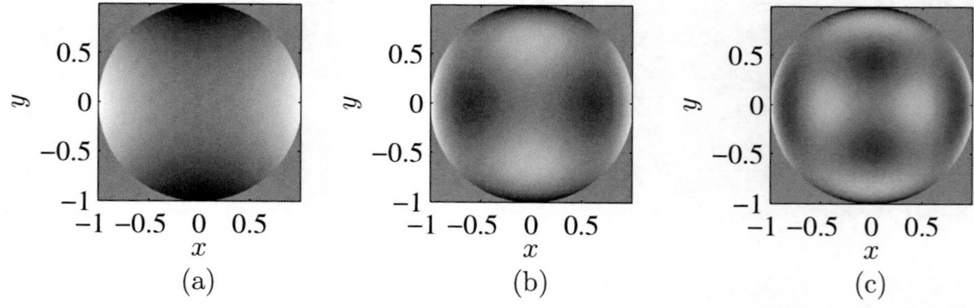

Figure 5.2 Plots of three orders of Zernike astigmatism. The wavefronts are shown for (a) $i = 6$, (b) $i = 12$, and (c) $i = 24$.

values of the wavefront and Zernike polynomials at the locations of x_p and y_q. To make this manifest, we define the notational changes

$$W_{pq} = W(x_p, y_q), \quad Z_{i,pq} = Z_i(x_p, y_q) \tag{5.12}$$

and use this new notation. This yields

$$a_i = \frac{\sum_p \sum_q W_{pq} Z_{i,pq}}{\sum_p \sum_q Z_{i,pq}^2}. \tag{5.13}$$

This notation can be simplified further by using only one index j to take the place of p and q. This means of referring to all wavefront and Zernike values within the aperture could be done in column-major, row-major, or any other order. The choice does not matter; however different programming (or scripting) languages handle certain orderings naturally. For example, C and C++ use row-major order, while MATLAB uses column-major order. Now, using just the index j for the different samples in the aperture gives

$$a_i = \frac{\sum_j W_j Z_{i,j}}{\sum_j Z_{i,j}^2}. \tag{5.14}$$

The same discretization and linear indexing could be applied to Eq. (5.6), leading to

$$W_j \cong \sum_{i=1}^{n_Z} Z_{i,j}\, a_i, \tag{5.15}$$

where n_Z is the number of modes being used. The reader should beware that the relationship is only approximate because of the discrete grid. The accuracy improves as more grid points are used.[24] This linear indexing now provides a new interpretation. We can treat Eq. (5.15) as a vector-matrix multiplication. Now, denote $\mathbf{W}$ as a column vector with elements W_i, Z as a matrix with elements Z_{ij}, and $\mathbf{A}$ as a column vector with elements A_i. To be explicit, the columns of Z are formed from individual Zernike polynomials evaluated at each aperture location such that

$$\mathsf{Z} = [Z_1 | Z_2 | \ldots | Z_{n_Z}], \tag{5.16}$$

where Z_1, Z_2, etc. are linear-indexed Zernike values. The number of rows in $\mathbf{W}$ is equal to the number of grid points within the aperture. The number of rows in $\mathbf{A}$ is equal to the number of modes being used. Correspondingly, the number of rows in Z is equal to the number of grid points, and the number of columns is equal to the number of modes. Finally, Eq. (5.15) compactly becomes

$$\mathbf{W} = \mathsf{Z}\mathbf{A}. \tag{5.17}$$

Listing 5.2 An example of computing Zernike coefficients from an arbitrary wavefront.

```
1  % example_zernike_projection.m
2
3  N = 32;     % number of grid points per side
4  L = 2;       % total size of the grid [m]
5  delta = L / N;  % grid spacing [m]
6  % cartesian & polar coordinates
7  [x y] = meshgrid((-N/2 : N/2-1) * delta);
8  [theta r] = cart2pol(x, y);
9  % unit circle aperture
10 ap = circ(x, y, 2);
11 % 3 Zernike modes
12 z2 = zernike(2, r, theta) .* ap;
13 z4 = zernike(4, r, theta) .* ap;
14 z21 = zernike(21, r, theta) .* ap;
15 % create the aberration
16 W = 0.5 *  z2 + 0.25 * z4 - 0.6 * z21;
17 % find only grid points within the aperture
18 idx = logical(ap);
19 % perform linear indexing in column-major order
20 W = W(idx);
21 Z = [z2(idx) z4(idx) z21(idx)];
22 % solve the system of equations to compute coefficients
23 A = Z \ W
```

Those familiar with linear algebra might recognize Eq. (5.14) as the Moore-Penrose pseudo-inverse (least-squares) solution to Eq. (5.17), written here in matrix notation as

$$\mathbf{A} = \left(\mathsf{Z}^T \mathsf{Z}\right)^{-1} \mathsf{Z}^T \mathbf{W}. \tag{5.18}$$

The vector-matrix forms here are compact in notation, and they can be implemented as a single line of code in many programming languages. For example, linear-algebra packages such as Linear Algebra PACKage (more commonly known as LAPACK)[25] and Basic Linear Algebra Subroutines (more commonly known as BLAS)[26,27], available for the C and FORTRAN languages, provide many fast-executing manipulations of matrices and vectors. Listing 5.2 gives a MATLAB example of projecting a complicated phase onto Zernike modes. The phase tested in the code is a weighted sum of modes 2, 4, and 21 with weights 0.5, 0.25, and -0.6, respectively. When the code is executed, the values in the array `A` are computed to be 0.5, 0.25, and -0.6, respectively.

5.1.2.1 Decomposition and mode removal

The previous subsection demonstrated how to compute the Zernike mode content of a phase map, given by its Zernike coefficients. Knowing this Zernike content can be quite useful. For example, we might have an optical system's measured aberration and wish to see what happens if we design an element to compensate for part of that aberration. As a practical instance, eye glasses and contact lenses often compensate for focus and astigmatism.

A real aberration $W(r,\theta)$ might contain a very large number of modes, but we may be interested in a mode-limited version $W'(r,\theta)$. Let us define

$$W'(r,\theta) = \sum_{i=1}^{n_Z} a_i Z_i(r,\theta) \tag{5.19}$$

as the mode-limited version of $W(r,\theta)$ such that

$$W(r,\theta) = W'(r,\theta) + \sum_{i=n_Z+1}^{\infty} a_i Z_i(r,\theta). \tag{5.20}$$

This is a good framework for partially corrected aberrations. With eye glasses and contact lenses, we ignore modes 1–3 because they do not affect visual image quality. Corrective lenses might compensate modes 4, 5, and 6. In that case, $n_Z = 6$, and the eyes see images blurred by the residual aberration containing modes $i = 7$ and up. Fortunately, the coefficients for these residual modes are usually much smaller than for the compensated modes.

An adaptive optics system is like a dynamically reconfigurable, high-resolution "contact lens" for imaging telescopes and cameras. A wavefront sensor is used to sense aberrations rapidly (sometimes over 10,000 frames per second) and adjust the figure of a deformable mirror to compensate aberrations.[23] Many of today's astronomical telescopes use adaptive optics to compensate phase aberrations caused by imaging through Earth's turbulent atmosphere. Deformable mirrors can only reproduce a finite number of Zernike modes, so there is always some residual aberration uncorrected by the mirror. Listing 5.3 gives an example of generating a random draw of a turbulent aberration and producing a mode-limited version $W'(r,\theta)$ (generating the aberration is covered in Sec. 9.3). Figure 5.3 shows the original screen and versions limited to 3, 16, 36, and 100 modes. Notice how the mode-limited version increasingly resembles the original aberration as more modes are included in the Zernike series representation.

It is also interesting to examine the residual phase of mode-limited aberrations. Figure 5.4 shows the complement [remaining terms, i.e., the second term in Eq. (5.20)] to each of Fig. 5.3's mode-limited aberrations. Notice how the structures in the residual phase get finer as more modes are included in the Zernike series representation. Also, note that adaptive optics systems typically use a fast steering mirror to compensate turbulence-induced tilt, leaving modes 4 and higher to be compensated by the deformable mirror. Accordingly, the residual phase in the upper left corner of Figure 5.4 shows the aberration that the deformable mirror must compensate. For a deformable mirror that can represent up to the first 100 Zernike modes, the lower right corner of Figure 5.4 shows the residual aberration after the deformable mirror that still blurs the image. As one can see in the figure, if adaptive optics are designed properly, it usually reduces the aberration significantly

Listing 5.3 An example of synthesizing a mode-limited version of an arbitrary aberration. The aberration in this example is a random draw of an atmospheric phase screen, discussed in Sec. 9.3.

```
1  % example_zernike_synthesis.m
2
3  N = 40;        % number of grid points per side
4  L = 2;         % total size of the grid [m]
5  delta = L / N;   % grid spacing [m]
6  % cartesian & polar coordinates
7  [x y] = meshgrid((-N/2 : N/2-1) * delta);
8  [theta r] = cart2pol(x, y);
9  % unit circle aperture
10 ap = circ(x, y, 2);
11 % indices of grid points in aperture
12 idxAp = logical(ap);
13 % create atmospheric phase screen
14 r0 = L / 20;
15 screen = ft_phase_screen(r0, N, delta, inf, 0) ...
16     / (2*pi) .* ap;
17 W = screen(idxAp);    % perform linear indexing
18
19 %%% analyze screen
20 nModes = 100;    % number of Zernike modes
21 % create matrix of Zernike polynomial values
22 Z = zeros(numel(W), nModes);
23 for idx = 1 : nModes
24     temp = zernike(idx, r, theta);
25     Z(:,idx) = temp(idxAp);
26 end
27 % compute mode coefficients
28 A = Z \ W;
29 % synthesize mode-limited screen
30 W_prime = Z*A;
31 % reshape mode-limited screen into 2-D for display
32 scr = zeros(N);
33 scr(idxAp) = W_prime;
```

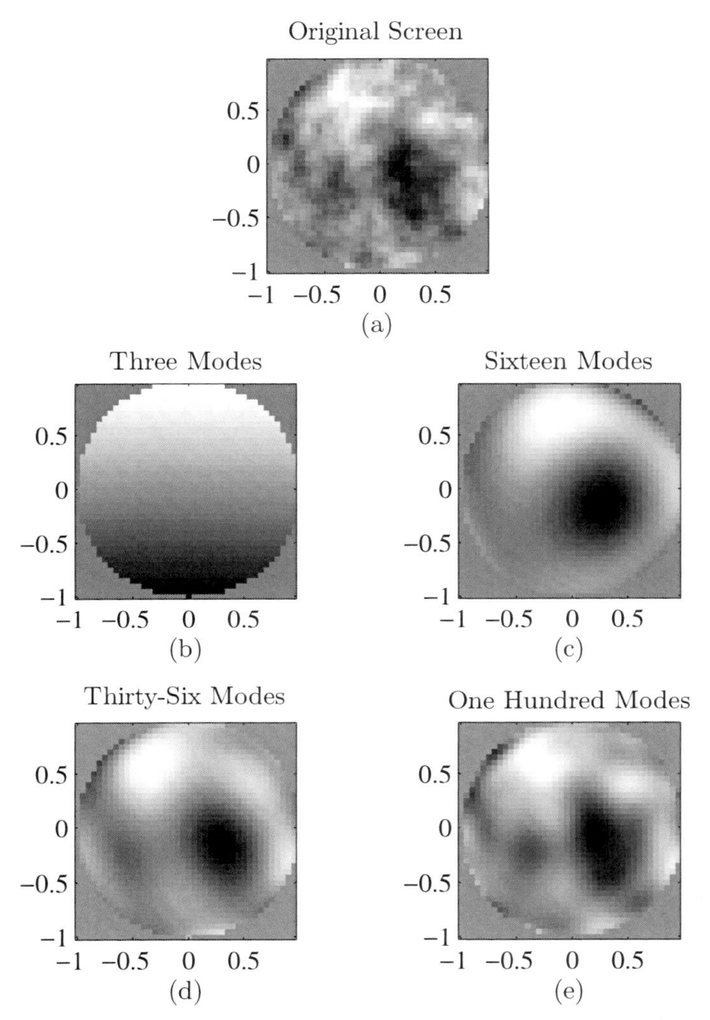

Figure 5.3 Plots of mode-limited phase screens. The original screen is at the top in plot (a). The four lower plots, (b)–(e) show the screen limited to 3, 16, 36, and 100 modes, respectively.

and provides greatly improved imagery.

5.1.2.2 RMS wavefront aberration

It is often handy to describe a wavefront aberration by its rms value σ averaged over the aperture. We compute the mean-square wavefront deviation straightforwardly via

$$\sigma^2 = \frac{1}{\pi} \int_0^{2\pi} \int_0^1 \left[W(r,\theta) - \overline{W}\right]^2 r\, dr\, d\theta, \tag{5.21}$$

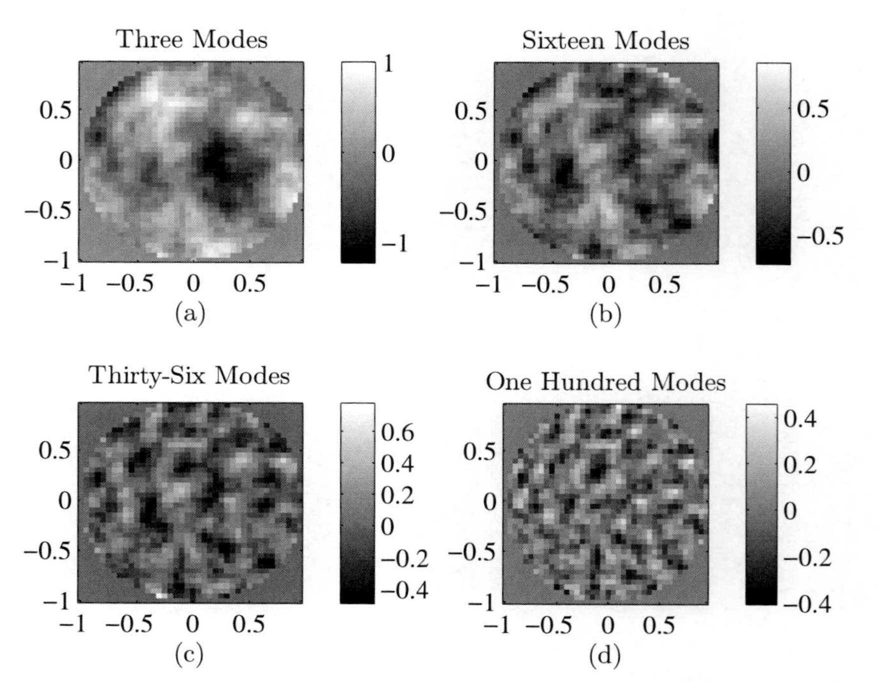

Figure 5.4 Plots of residual phase due to finite number of modes. These are the residuals for the mode limits in Fig. 5.3.

where $\overline{W}$ is the mean of W over the aperture. Note that in Eq. (5.21), the average is over the pupil area, which is π for a unit-radius circle. Writing the wavefront as a Zernike series yields

$$\sigma^2 = \frac{1}{\pi} \int_0^{2\pi} \int_0^1 \left[\sum_{i=2}^{\infty} a_i Z_i(r,\theta) \right]^2 r\, dr\, d\theta, \tag{5.22}$$

where the reader should note that the sum begins at $i = 2$ because $\overline{W}$ is the $i = 1$ term. We now factor the squared sum into an explicit product of two series so that

$$\sigma^2 = \frac{1}{\pi} \int_0^{2\pi} \int_0^1 \left[\sum_{i=2}^{\infty} a_i Z_i(r,\theta) \right] \left[\sum_{i'=2}^{\infty} a_{i'} Z_{i'}(r,\theta) \right] r\, dr\, d\theta \tag{5.23}$$

$$= \frac{1}{\pi} \sum_{i=2}^{\infty} a_i \sum_{i'=2}^{\infty} a_{i'} \int_0^{2\pi} \int_0^1 Z_i(r,\theta)\, Z_{i'}(r,\theta)\, r\, dr\, d\theta \tag{5.24}$$

$$= \frac{1}{\pi} \sum_{i=2}^{\infty} a_i \sum_{i'=2}^{\infty} a_{i'}\, \pi \delta_{ii'} \tag{5.25}$$

$$= \sum_{i=2}^{\infty} a_i^2. \tag{5.26}$$

This means that the wavefront variance can be found by simply summing the squares of the Zernike coefficients. This is a very convenient benefit of using an orthogonal basis set to describe aberrations.

5.2 Impulse Response and Transfer Function of Imaging Systems

Aberrations have a strong effect on the impulse response of an imaging system. Further, the imaging system model shown in Fig. 5.1 has different impulse responses depending on the coherence of the object's illumination. If the illumination is spatially coherent, the impulse response is called the amplitude spread function (or coherent spread function), and the system's frequency response is called the amplitude transfer function (or coherent transfer function).[5] This is discussed in Sec. 5.2.1. If the illumination is spatially incoherent, the impulse response is called the point spread function, and the system's frequency response is called the optical transfer function (OTF), and its magnitude is called the modulation transfer function (MTF). This is discussed in Sec. 5.2.2.

Note that wavefront aberrations are independent of the illumination. They only depend on the optical components of the imaging system. However, their effect on the image does depend on the coherence of the illumination.

5.2.1 Coherent imaging

When the light is coherent, imaging systems are linear in optical field. Accordingly, the image amplitude $U_i(u,v)$ is the convolution of the object amplitude $U_o(u,v)$ with the amplitude spread function $h(u,v)$ according to

$$U_i(u,v) = \int_{-\infty}^{\infty}\int_{-\infty}^{\infty} h(u-\eta, v-\xi)\, U_o(\eta,\xi)\; d\xi\, d\eta \tag{5.27}$$

$$= h(u,v) \otimes U_o(u,v)\,. \tag{5.28}$$

This assumes that the imaging system has unit magnification. Accounting for magnification just requires scaling of the object coordinates.[5] The amplitude spread function is given by

$$h(u,v) = \frac{1}{\lambda z_i}\int_{-\infty}^{\infty}\int_{-\infty}^{\infty} \mathcal{P}(x,y)\, e^{-i\frac{2\pi}{\lambda z_i}(ux+vy)}\, dx\, dy \tag{5.29}$$

$$= \frac{1}{\lambda z_i}\mathcal{F}\left\{\mathcal{P}(x,y)\right\}_{f_x=\frac{u}{\lambda z_i},\, f_y=\frac{v}{\lambda z_i}}\,, \tag{5.30}$$

where $\mathcal{P}(x,y)$ is the generalized pupil function defined in Eq. (5.1) and z_i is the image distance.

Listing 5.4 An example of coherent imaging in MATLAB.

```
1  % example_coh_img.m
2
3  N = 256;      % number of grid points per side
4  L = 0.1;         % total size of the grid [m]
5  D = 0.07;     % diameter of pupil [m]
6  delta = L / N;   % grid spacing [m]
7  wvl = 1e-6; % optical wavelength [m]
8  z = 0.25;     % image distance [m]
9  % pupil-plane coordinates
10 [x y] = meshgrid((-N/2 : N/2-1) * delta);
11 [theta r] = cart2pol(x, y);
12 % wavefront aberration
13 W = 0.05 * zernike(4, 2*r/D, theta);
14 % complex pupil function
15 P = circ(x, y, D) .* exp(i * 2*pi * W);
16 % amplitude spread function
17 h = ft2(P, delta);
18 delta_u = wvl * z / (N*delta);
19 % image-plane coordinates
20 [u v] = meshgrid((-N/2 : N/2-1) * delta_u);
21 % object (same coordinates as h)
22 obj = (rect((u-1.4e-4)/5e-5) + rect(u/5e-5) ...
23     + rect((u+1.4e-4)/5e-5)) .* rect(v/2e-4);
24 % convolve the object with the ASF to simulate imaging
25 img = myconv2(obj, h, 1);
```

Listing 5.4 gives an example of how to compute a coherent image given the object and amplitude spread function of the imaging system. In the example, the object comprises three parallel rectangular slits as shown in Fig. 5.5(a). The aberration is 0.05 waves of Zernike defocus ($i = 4$), computed in line 13. The resulting generalized pupil function is computed in line 15. Line 17 computes the amplitude spread function using the `ft2` function, and it is shown in Fig. 5.5(b). Notice that is much narrower than the object. As noted in Sec. 3.1, this is typical of impulse responses in linear systems. Finally, the image field is formed by convolving the object field and amplitude spread function in line 25 using the `conv2` function. The resulting object intensity is shown in Fig. 5.5.

If the convolution theorem is applied to Eq. (5.27), the result is

$$\mathcal{F}\left\{U_i\left(u,v\right)\right\} = \mathcal{F}\left\{h\left(u,v\right)\right\}\mathcal{F}\left\{U_o\left(u,v\right)\right\}. \tag{5.31}$$

In this form, it is clear that the amplitude spread function's Fourier spectrum modulates the object's spectrum to yield the the diffraction image. This specifies how object's frequency spectrum is transferred through the imaging system to the diffrac-

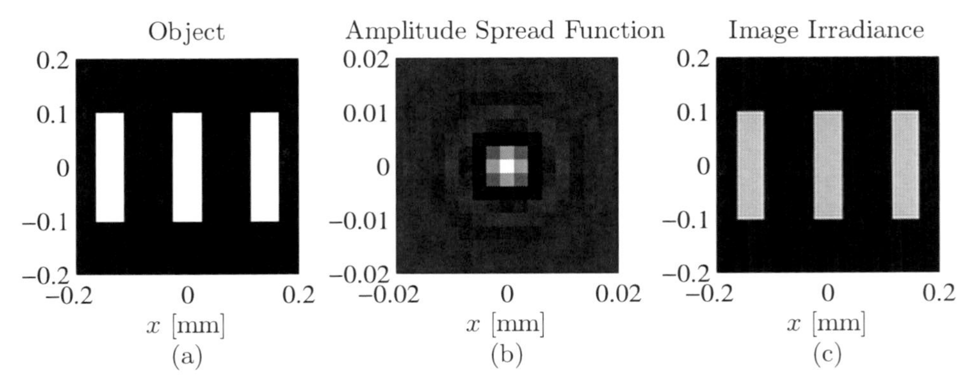

Figure 5.5 Example of coherent imaging. Plot (a) shows the object, while plot (b) shows the amplitude spread function due to defocus, and plot (c) shows the coherent image blurred by 0.05 waves of defocus.

tion image, so we define this property of the system as the amplitude transfer function given by

$$H(f_x, f_y) = \mathcal{F}\{h(u, v)\} \tag{5.32}$$

$$= \mathcal{F}\left\{\frac{1}{\lambda z_i}\mathcal{F}\{\mathcal{P}(x, y)\}_{f_x=\frac{u}{\lambda z_i}, f_y=\frac{v}{\lambda z_i}}\right\} \tag{5.33}$$

$$= \lambda z_i \mathcal{P}(-\lambda z_i f_x, -\lambda z_i f_y). \tag{5.34}$$

In the last equation, Eq. (5.30) has been used to write the amplitude transfer function in terms of system's pupil function. The low-pass filtering property of imaging systems is now evident when we consider a common aperture like a circle. Eq. (5.34) indicates that a circular aperture with diameter D would pass all frequencies for which $\left(f_x^2 + f_y^2\right)^{1/2} < D/(2\lambda z_i)$ equally while filtering out all higher frequencies completely. In this way, image amplitude is a strictly bandlimited function.

5.2.2 Incoherent imaging

When the light is spatially incoherent, the image irradiance is the convolution of the object irradiance with the point spread function (PSF):

$$I_i(u, v) = \int_{-\infty}^{\infty}\int_{-\infty}^{\infty} |h(u - \eta, v - \xi)|^2 I(\eta, \xi)\, d\xi\, d\eta \tag{5.35}$$

$$= |h(u, v)|^2 \otimes I(u, v). \tag{5.36}$$

The point spread function is simply the squared magnitude of the amplitude spread function. Listing 5.5 gives an example of how to compute an incoherent image given the object and amplitude spread function of the imaging system. The

Listing 5.5 An example of incoherent imaging in MATLAB.

```
1  % example_incoh_img.m
2
3  N = 256;     % number of grid points per side
4  L = 0.1;        % total size of the grid [m]
5  D = 0.07;    % diameter of pupil [m]
6  delta = L / N;   % grid spacing [m]
7  wvl = 1e-6; % optical wavelength [m]
8  z = 0.25;    % image distance [m]
9  % pupil-plane coordinates
10 [x y] = meshgrid((-N/2 : N/2-1) * delta);
11 [theta r] = cart2pol(x, y);
12 % wavefront aberration
13 W = 0.05 * zernike(4, 2*r/D, theta);
14 % complex pupil function
15 P = circ(x, y, D) .* exp(i * 2*pi * W);
16 % amplitude spread function
17 h = ft2(P, delta);
18 U = wvl * z / (N*delta);
19 % image-plane coordinates
20 [u v] = meshgrid((-N/2 : N/2-1) * U);
21 % object (same coordinates as h)
22 obj = (rect((u-1.4e-4)/5e-5) + rect(u/5e-5) ...
23     + rect((u+1.4e-4)/5e-5)) .* rect(v/2e-4);
24 % convolve the object with the PSF to simulate imaging
25 img = myconv2(abs(obj).^2, abs(h).^2, 1);
```

object and aberration are the same as those from the coherent example. The basic computations are the same, too, except that the object irradiance is convolved with the imaging system's point spread function. The results are shown in Fig. 5.6.

Like the coherent case, the convolution theorem can be applied to Eq. (5.35), and now the result is

$$\mathcal{F}\left\{I_i\left(u,v\right)\right\} = \mathcal{F}\left\{\left|h\left(u,v\right)\right|^2\right\}\mathcal{F}\left\{I_o\left(u,v\right)\right\}. \tag{5.37}$$

Again, we can see that the PSF's Fourier spectrum modulates the object irradiance's spectrum to yield the diffraction image. In the incoherent case, the filter function (called the optical transfer function) is defined as

$$\mathcal{H}\left(f_x,f_y\right) = \frac{\mathcal{F}\left\{\left|h\left(u,v\right)\right|^2\right\}}{\int\limits_{-\infty}^{\infty}\int\limits_{-\infty}^{\infty}\left|h\left(u,v\right)\right|^2 du dv}. \tag{5.38}$$

Similarly to the coherent case, we can relate this to the pupil function. Application

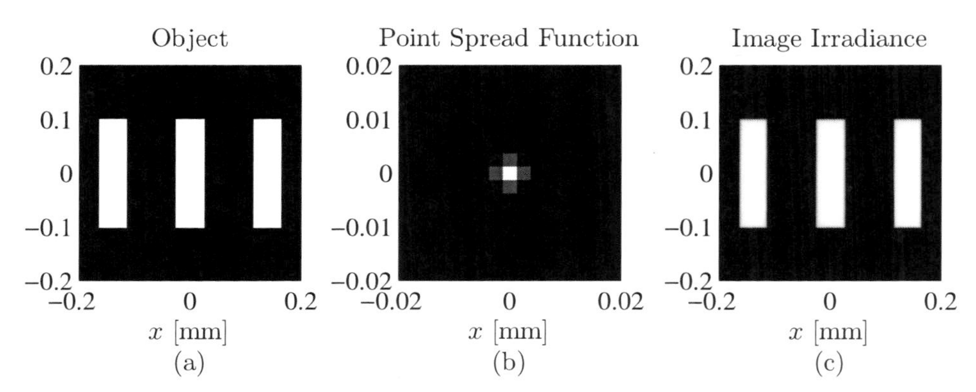

Figure 5.6 Example of incoherent imaging. Plot (a) shows the object, while plot (b) shows the point spread function due to defocus, and plot (c) shows the incoherent image blurred by 0.05 waves of defocus.

of the auto-correlation theorem and Parseval's theorem yields

$$\mathcal{H}(f_x, f_y) = \frac{\int\limits_{-\infty}^{\infty}\int\limits_{-\infty}^{\infty} H^*(p - f_x, q - f_y)\, H(p, q)\; dp\, dq}{\int\limits_{-\infty}^{\infty}\int\limits_{-\infty}^{\infty} |H(p, q)|^2\; dp\, dq} \tag{5.39}$$

$$= \frac{\int\limits_{-\infty}^{\infty}\int\limits_{-\infty}^{\infty} \mathcal{P}^*(x - \lambda z_i f_x, y - \lambda z_i f_y)\, \mathcal{P}(x, y)\; dx\, dy}{\int\limits_{-\infty}^{\infty}\int\limits_{-\infty}^{\infty} |\mathcal{P}(x, y)|^2\; dx\, dy} \tag{5.40}$$

$$= \left.\frac{\mathcal{P}^*(x, y) \star \mathcal{P}(x, y)}{\int\limits_{-\infty}^{\infty}\int\limits_{-\infty}^{\infty} |\mathcal{P}(x, y)|^2\; dx\, dy}\right|_{x=\lambda z_i f_x, y=\lambda z_i f_y}. \tag{5.41}$$

The example case of a circular aperture with diameter D is illustrative again. It can be shown that the OTF for a circular aperture is an azimuthally symmetric function of $f = \left(f_x^2 + f_y^2\right)^{1/2}$ given by

$$\mathcal{H}(f) = \begin{cases} \frac{2}{\pi}\left[\cos^{-1}\left(\frac{f}{2f_0}\right) - \frac{f}{2f_0}\sqrt{1 - \left(\frac{f}{2f_0}\right)^2}\right] & f \leq 2f_0 \\ 0 & \text{otherwise,} \end{cases} \tag{5.42}$$

where $f_0 = D/(2\lambda z_i)$. This quantity f_0 is the cutoff frequency for the coherent case, but as Eq. (5.42) indicates, frequencies up to $2f_0$ pass through (with some attenuation) when the light is incoherent. Still, incoherent images are strictly bandlimited. Another difference from the coherent case is that $\mathcal{H}(f) \geq 0$ for all frequencies.

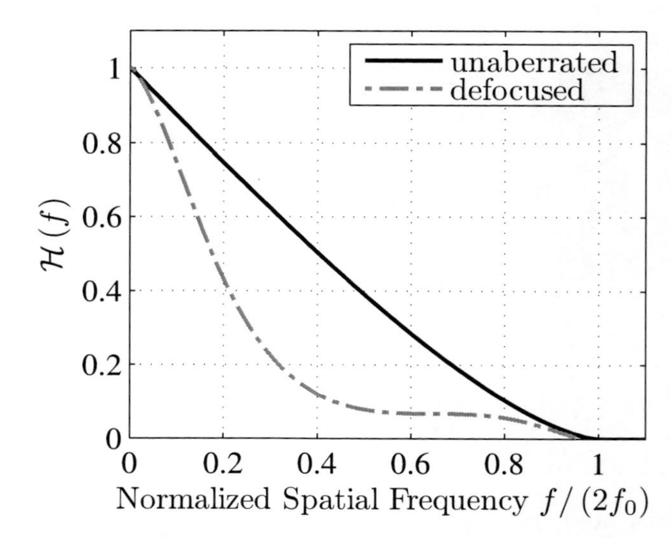

Figure 5.7 Optical transfer functions for unaberrated and defocused imaging systems.

Figure 5.7 shows a plot of two OTFs for imaging systems with circular apertures. The solid black line is the OTF for a system without aberrations as given in Eq. (5.42). The dash-dot gray line is the OTF for a system with defocus such that the wavefront error is 0.5 waves at the edge of the aperture (computed by numerical integration). Clearly, the defocused image would have many frequency components that are more attenuated than an aberration-free image. This is also characterized by a broader PSF, and results in a blurred image. The next subsection discusses a related metric for image quality.

5.2.3 Strehl ratio

Clearly, the performance of an imaging system is determined by its amplitude/-point spread function. It is handy to have a single-number metric to describe performance. The most common metric is Strehl ratio, which is the ratio of the on-axis actual point spread function value to the on-axis ideal point spread function value. Typically, this is a comparison of an aberrated system to an almost identical but unaberrated system. The on-axis value of a PSF is computed by using Eq. (5.29) at the origin:

$$|h(0,0)|^2 = \frac{1}{\lambda^2 z_i^2} \left| \int_{-\infty}^{\infty} \int_{-\infty}^{\infty} \mathcal{P}(x,y)\, e^0\, dx\, dy \right|^2 \tag{5.43}$$

$$= \frac{1}{\lambda^2 z_i^2} \left| \int_{-\infty}^{\infty} \int_{-\infty}^{\infty} \mathcal{P}(x,y)\, dx\, dy \right|^2 . \tag{5.44}$$

Because the only contribution to non-zero phase in the generalized pupil function $\mathcal{P}(x, y)$ is caused by aberrations, $P(x, y)$ is the unaberrated pupil function. As a result, the Strehl ratio $\mathcal{S}$ is computed as

$$\mathcal{S} = \frac{\left| \int_{-\infty}^{\infty} \int_{-\infty}^{\infty} \mathcal{P}(x, y)\, dx\, dy \right|^2}{\left| \int_{-\infty}^{\infty} \int_{-\infty}^{\infty} P(x, y)\, dx\, dy \right|^2}. \tag{5.45}$$

To make the aberration phase $\phi(x, y)$ more manifest, we can rewrite Eq. (5.45) as

$$\mathcal{S} = \frac{\left| \int_{-\infty}^{\infty} \int_{-\infty}^{\infty} P(x, y)\, e^{i\phi(x,y)}\, dx\, dy \right|^2}{\left| \int_{-\infty}^{\infty} \int_{-\infty}^{\infty} P(x, y)\, dx\, dy \right|^2} \tag{5.46}$$

$$= \frac{\int_{-\infty}^{\infty} \mathcal{H}(f_x, f_y)\, df_x\, df_y}{\int_{-\infty}^{\infty} \mathcal{H}_{dl}(f_x, f_y)\, df_x\, df_y}, \tag{5.47}$$

where Eqs. (5.30) and (5.38) have been applied to obtain the latter equation and $\mathcal{H}_{dl}(f_x, f_y)$ is the OTF of an unaberrated (or diffraction-limited) system.

For a perfectly unaberrated system, $\mathcal{S} = 1$, and this is the maximum possible value of the Strehl ratio. Aberrations and amplitude variations in the pupil (for example, an annular aperture) always reduce the Strehl ratio.[19] Consequently, low Strehl ratio indicates poor image quality, i.e, coarse resolution and low contrast.

For small aberrations, the Strehl ratio of an image is determined by the variance of the pupil phase. To show this, we can rewrite Eq. (5.46) in the abbreviated form

$$\mathcal{S} = \left| \left\langle e^{i\phi} \right\rangle \right|^2, \tag{5.48}$$

where the angle brackets $\langle \ldots \rangle$ indicate a spatial average over the amplitude-weighted pupil. For example, the amplitude-weighted average phase is given by[19]

$$\langle \phi \rangle = \frac{\int_{-\infty}^{\infty} \int_{-\infty}^{\infty} P(x, y)\, \phi(x, y)\, dx\, dy}{\int_{-\infty}^{\infty} \int_{-\infty}^{\infty} P(x, y)\, dx\, dy}. \tag{5.49}$$

Multiplying Eq. (5.48) by $\left| e^{-i\langle \phi \rangle} \right|^2 = 1$ yields

$$\mathcal{S} = \left| \left\langle e^{i(\phi - \langle \phi \rangle)} \right\rangle \right|^2 \tag{5.50}$$

$$= \langle \cos(\phi - \langle\phi\rangle) \rangle^2 + \langle \sin(\phi - \langle\phi\rangle) \rangle^2 . \tag{5.51}$$

Taking the first terms up to second order of the Taylor-series expansions gives

$$\mathcal{S} \simeq \left\langle 1 - \frac{(\phi - \langle\phi\rangle)^2}{2} \right\rangle^2 + \langle \phi - \langle\phi\rangle \rangle^2 \tag{5.52}$$

$$\simeq \left(1 - \frac{\sigma_\phi^2}{2}\right)^2 . \tag{5.53}$$

Carrying out the multiplication and keeping only the first two terms leads to

$$\mathcal{S} \simeq 1 - \sigma_\phi^2, \tag{5.54}$$

where $\sigma_\phi^2 = 4\pi^2\sigma^2$ is the variance of the phase, measured in rad^2. This result is the same as writing

$$\mathcal{S} \simeq e^{-\sigma_\phi^2} \tag{5.55}$$

and keeping only the first two terms in its Taylor series expansion. Eqs. (5.53)–(5.55) all represent commonly used approximations for computing Strehl ratio. Eq. (5.53) is the Maréchal formula. Eq. (5.55), while it is presented here as an approximation to Eq. (5.54), actually is an empirical formula that gives the best fit to numerical results for various aberrations.[19]

5.3 Problems

1. The Sellmeier equation is an empirical relationship between optical wavelength and refractive index for glass. It is given by

$$n^2(\lambda) = 1 + \sum_i \frac{B_i \lambda^2}{\lambda^2 - C_i} \tag{5.56}$$

   Each type of glass has its own measured set of Sellmeier coefficients B_i and C_i.

   (a) Find the Sellmeier coefficients for borosilicate crown glass (more commonly called BK7) and compute the standard refractive indices

$$n_F = n(486.12\,\text{nm}) \quad \text{blue Hydrogen line} \tag{5.57}$$

$$n_d = n(587.56\,\text{nm}) \quad \text{yellow Helium line} \tag{5.58}$$

$$n_C = n(656.27\,\text{nm}) \quad \text{red Hydrogen line} \tag{5.59}$$

   to six significant digits.

(b) You are given a thin plano-convex lens made of BK7 glass. The convex side is spherical with a 51.68-mm radius of curvature, and the lens diameter is 12.7 mm. Compute the focal lengths and diffraction-limited spot diameters corresponding to each of the standard wavelengths from part (a).

(c) Follow the coherent-imaging example of Sec. 5.2.1 to compute each diffraction-limited PSF. Add several different levels of defocus aberration and compute the resulting PSFs. For all wavelengths, plot the $v = 0$ slice of each PSF to demonstrate how the focal spot evolves near the geometric focal plane. Use these PSF-slice plots to show that you have computed the correct spot diameters. Use 1024 grid points per side and a grid spacing of 0.199 mm.

2. For a lens that is aberrated with one wave of Zernike primary astigmatism, add several different levels of defocus aberration and compute the resulting PSFs. Show images of these PSFs to demonstrate how the focal spot evolves near the geometric focal plane. Use a grid size = 4 m, aperture diameter = 2 m, with 512 points per side, optical wavelength $= 1\mu$m, and focal length $= 16$ m.

3. For a lens that is aberrated with one wave of Zernike primary spherical aberration, add several different levels of defocus aberration and compute the resulting PSFs. Show images of these PSFs to demonstrate how the focal spot evolves near the geometric focal plane. Use a grid size $= 4$ m, aperture diameter $= 2$ m, with $= 512$ points per side, optical wavelength $= 1\mu$m, and focal length $= 16$ m.

4. Given

$$W(x, y) = 0.07\, Z_4 + 0.05\, Z_5 - 0.05\, Z_6 + 0.03\, Z_7 - 0.03\, Z_8, \qquad (5.60)$$

compute the Strehl ratio

(a) using Eqs. (5.26) and (5.55),

(b) and using a simulation to compute the aberrated and diffraction-limited PSFs (similar to the example of Sec. 5.2.1). Use a grid size $= 8$ m, aperture diameter $= 2$ m, with $= 512$ points per side, optical wavelength $= 1\mu$m, and focal length $= 64$ m.

5. Numerically compute the PSF of an annular aperture whose inner and outer diameters are 1 m and 2 m, respectively. Also compute the PSF of a filled 2 m circular aperture. Use a grid size $= 8$ m, with $= 512$ points per side, optical wavelength $= 1\mu$m, and focal length $= 64$ m. Provide displays of both PSFs and compute the Strehl ratio of the annular aperture as the ratio of the peaks of the PSFs. Confirm your numerical results with analytic calculations.

6. Numerically compute the PSF of a sparse (or aggregate) aperture composed of three 1-m-diameter circular apertures each centered at coordinates $(0.6, 0.6)$ m, $(-0.6, 0.6)$ m, and $(0, 0.6)$ m. Use a grid size $=$ 8 m, grid size $=$ 512 points per side, optical wavelength $= 1\mu$m, and focal length $=$ 64 m. Provide displays of the aperture and PSF. Confirm your numerical results with analytic calculations.

Chapter 6
Fresnel Diffraction in Vacuum

The goal of this chapter is to develop methods for modeling near-field optical-wave propagation with high fidelity and some flexibility, which is considerably more challenging than for far-field propagation. This chapter uses the same coordinate convention as in Fig. 1.2. It begins with a discussion of different forms of the Fresnel diffraction integral. These different forms can be numerically evaluated in different ways, each with benefits and drawbacks. Then, to emphasize the different mathematical operations in the notation, operators are introduced that are used throughout Chs. 6–8. The rest of this chapter develops basic algorithms for wave propagation in vacuum and other simulation details.

The quadratic phase factor inside the Fresnel diffraction integral is not bandlimited, so it poses some challenges related to sampling. There are two different ways to evaluate the integral: as a single FT or as a convolution. This chapter develops both basic methods as well as more sophisticated versions that provide some flexibility. There are different types of flexibility that one might need. For example, Delen and Hooker present a method that is particularly useful for simulating propagation in integrated optical components. Because the interfaces in these components are often slanted or offset and the angles are not always paraxial, they developed a Rayleigh-Summerfeld propagation method that can handle propagation between arbitrarily oriented planes with good accuracy.[28,29]

In contrast, the applications discussed in this book involve parallel source and observation planes, and the paraxial approximation is a very good one. When long propagation distances are involved, beams can spread to be much larger than their original size. Accordingly, some algorithms discussed in this chapter provide the user with the flexibility to choose the scaling between the observation- and source-plane grid spacings. Many authors have presented algorithms with this ability including Tyler and Fried,[30] Roberts,[31] Coles,[32] Rubio,[33] Deng *et al.*,[34] Coy,[35] Rydberg and Bengtsson,[36] and Voelz and Roggemann.[37] Most of these methods are mathematically equivalent to each other. However, one unique algorithm was presented by Coles[32] and later augmented by Rubio[33] in which a diverging spherical coordinate system was used by an angular grid with constant angular grid spacing. This was done specifically because the source was a point source, which naturally diverges spherically. Rubio augmented this basic concept to allow for very long propagation distances. When the grid grows too large to adequately sample the field, Rubio's method is to extract a central portion and interpolate it to a finer grid.

In this chapter, two flexible propagation methods are presented. The first uses two steps of evaluating the Fresnel diffraction integral, with the grid spacings adjusted by the distances of the two propagations. The second method uses some algebraic manipulation of the convolution form of the Fresnel diffraction integral. The manipulation introduces a free parameter that directly sets the observation-plane grid spacing.

6.1 Different Forms of the Fresnel Diffraction Integral

We start with the Fresnel diffraction integral, which is repeated here for convenience:

$$U(x_2, y_2) = \frac{e^{ik\Delta z}}{i\lambda\Delta z} \int_{-\infty}^{\infty}\int_{-\infty}^{\infty} U(x_1, y_1)\, e^{i\frac{k}{2\Delta z}\left[(x_2-x_1)^2+(y_2-y_1)^2\right]}\, dx_1\, dy_1. \quad (6.1)$$

Also, we define spatial and spatial-frequency vectors

$$\mathbf{r}_1 = x_1\hat{\mathbf{i}} + y_1\hat{\mathbf{j}} \quad (6.2)$$

$$\mathbf{r}_2 = x_2\hat{\mathbf{i}} + y_2\hat{\mathbf{j}} \quad (6.3)$$

$$\mathbf{f}_1 = f_{x1}\hat{\mathbf{i}} + f_{y1}\hat{\mathbf{j}}, \quad (6.4)$$

where $\mathbf{r}_1$ is in the source plane, and $\mathbf{r}_2$ is in the observation plane. This is used throughout the chapter. Table 6.1 summarizes these quantities and others that are important to this development.

We want to use the Fresnel diffraction integral to compute the observation-plane optical field from knowledge of the source-plane field. Sections 6.3 and 6.4 deal with numerically evaluating this equation. There are two forms of Eq. (6.1) that are used for numerical evaluation. The first comes about by expanding the squared terms in the exponential and factoring portions out of the integral. This yields

$$U(x_2, y_2) = \frac{e^{ik\Delta z}}{i\lambda\Delta z}\, e^{i\frac{k}{2\Delta z}\left(x_2^2+y_2^2\right)} \times \int_{-\infty}^{\infty}\int_{-\infty}^{\infty} U(x_1, y_1)\, e^{i\frac{k}{2\Delta z}\left(x_1^2+y_1^2\right)} e^{-i\frac{2\pi}{\lambda\Delta z}\left(x_2x_1+y_2y_1\right)}\, dx_1\, dy_1, \quad (6.5)$$

which can be evaluated as an FT as discussed in Sec. 6.3. The second form of Eq. (6.1) comes about by noting that it is a convolution of the source-plane field with the free-space amplitude spread function so that

$$U(x_2, y_2) = U(x_1, y_1) \otimes \left[\frac{e^{ik\Delta z}}{i\lambda\Delta z}\, e^{i\frac{k}{2\Delta z}\left(x_1^2+y_1^2\right)}\right]. \quad (6.6)$$

Then, the convolution theorem is used to evaluate Eq. (6.6) via two FTs.

Table 6.1 Definition of symbols for Fresnel propagation.

symbol	meaning
$\mathbf{r}_1 = (x_1, y_1)$	source-plane coordinates
$\mathbf{r}_2 = (x_2, y_2)$	observation-plane coordinates
δ_1	grid spacing in source plane
δ_2	grid spacing in observation plane
$\mathbf{f}_1 = (f_{x1}, f_{y1})$	spatial-frequency of source plane
δ_{f1}	grid spacing in source-plane spatial frequency
z_1	location of source plane along the optical axis
z_2	location of observation plane along the optical axis
Δz	distance between source plane and observation plane
m	scaling factor from source plane to observation plane

6.2 Operator Notation

Operator notation is useful in Fresnel diffraction computations for writing the equations compactly without explicit integral notation. Using operators places the emphasis on operations that are taking place. The notation used here is adapted from that described by Nazarathy and Shamir,[38] who also incorporated it with ray matrices to describe diffraction through optical systems.[39] The key difference is that we specify the domains in which they operate. These operators are defined by:

$$\mathcal{Q}[c, \mathbf{r}]\{U(\mathbf{r})\} \equiv e^{i\frac{k}{2}c|\mathbf{r}|^2} U(\mathbf{r}) \tag{6.7}$$

$$\mathcal{V}[b, \mathbf{r}]\{U(\mathbf{r})\} \equiv b\, U(b\mathbf{r}) \tag{6.8}$$

$$\mathcal{F}[\mathbf{r}, \mathbf{f}]\{U(\mathbf{r})\} \equiv \int_{-\infty}^{\infty} U(\mathbf{r})\, e^{-i2\pi\mathbf{f}\cdot\mathbf{r}}\, d\mathbf{r} \tag{6.9}$$

$$\mathcal{F}^{-1}[\mathbf{f}, \mathbf{r}]\{U(\mathbf{f})\} \equiv \int_{-\infty}^{\infty} U(\mathbf{f})\, e^{i2\pi\mathbf{f}\cdot\mathbf{r}}\, d\mathbf{f} \tag{6.10}$$

$$\mathcal{R}[d, \mathbf{r}_1, \mathbf{r}_2]\{U(\mathbf{r}_1)\} \equiv \frac{1}{i\lambda d} \int_{-\infty}^{\infty} U(\mathbf{r}_1)\, e^{i\frac{k}{2d}|\mathbf{r}_2-\mathbf{r}_1|^2}\, d\mathbf{r}_1. \tag{6.11}$$

The operators' parameters are given in square brackets, and the operand is given in curly braces. Note that in Eqs. (6.9) and (6.10), the domain of the operand is listed as the first parameter, and the domain of the result is listed as the second parameter. See Refs. 38 and 39 for relations betweens these operators. Finally, we define one more quadratic-phase exponential operator

$$\mathcal{Q}_2[d, \mathbf{r}]\{U(\mathbf{r})\} \equiv e^{i\pi^2\frac{2d}{k}|\mathbf{r}|^2} U(\mathbf{r}). \tag{6.12}$$

The operator $\mathcal{Q}_2[d, \mathbf{r}]$ is not defined by Nazarathy and Shamir. In fact, it can be written in terms of the operator $\mathcal{Q}$ as

$$\mathcal{Q}_2[d, \mathbf{r}] = \mathcal{Q}\left[\frac{4\pi^2}{k^2}d, \mathbf{r}\right]. \tag{6.13}$$

However, it is just a convenient definition for use in Sec. 6.4.

6.3 Fresnel-Integral Computation

This section describes two methods of implementing the Fresnel diffraction integral in the form of Eq. (6.5). The first method evaluates this integral once as a single FT, which is the most straightforward. This method is desirable because of its computational efficiency. The second method evaluates the Fresnel integral twice, which adds some flexibility in the grid spacing at the cost of performing a second FT.

6.3.1 One-step propagation

Figure 1.2 shows the geometry of propagation from the source plane to the observation plane. The Fresnel integral can be used via Eq. (6.5) to compute the observation-plane field $U(x_2, y_2)$ directly, given the source-plane field $U(x_1, y_1)$ and the propagation geometry. We write Eq. (6.5) in operator notation as

$$U(\mathbf{r}_2) = \mathcal{R}[\Delta z, \mathbf{r}_1, \mathbf{r}_2]\{U(\mathbf{r}_1)\} \tag{6.14}$$

$$= \mathcal{Q}\left[\frac{1}{\Delta z}, \mathbf{r}_2\right]\mathcal{V}\left[\frac{1}{\lambda\Delta z}, \mathbf{r}_2\right]\mathcal{F}[\mathbf{r}_1, \mathbf{f}_1]\,\mathcal{Q}\left[\frac{1}{\Delta z}, \mathbf{r}_1\right]\{U(\mathbf{r}_1)\}. \tag{6.15}$$

The order of operation is right to left. In general, these operators do not commute; only certain combinations commute. It is clear that the observation-plane field is computed by (reading right to left) multiplying the source field by a quadratic phase ($\mathcal{Q}$), Fourier transforming ($\mathcal{F}$), scaling by a constant [$\mathcal{V}$ transforms from spatial frequency to spatial coordinates with $\mathbf{f}_1 = \mathbf{r}_2/(\lambda\Delta z)$], and multiplying by another quadratic phase factor ($\mathcal{Q}$). An intuitive explanation is that propagation can be represented as an FT between confocal spheres centered at the source and observation planes. The spheres' common radius of curvature is Δz.

To evaluate the Fresnel integral on a computer, again we must use a sampled version of the source-plane optical field $U(\mathbf{r}_1)$. Let the spacing in the source plane be δ_1. As before, the spacing in the frequency domain is $\delta_{f1} = 1/(N\delta_1)$, so then the spacing in the observation plane is

$$\delta_2 = \frac{\lambda\Delta z}{N\delta_1}. \tag{6.16}$$

Listing 6.1 gives the MATLAB function `one_step_prop` that numerically evaluates Eq. (6.5).

Listing 6.1 Code for evaluating the Fresnel diffraction integral in MATLAB using a single step.

```
function [x2 y2 Uout] ...
    = one_step_prop(Uin, wvl, d1, Dz)
% function [x2 y2 Uout] ...
%     = one_step_prop(Uin, wvl, d1, Dz)

    N = size(Uin, 1);   % assume square grid
    k = 2*pi/wvl;     % optical wavevector
    % source-plane coordinates
    [x1 y1] = meshgrid((-N/2 : 1 : N/2 - 1) * d1);
    % observation-plane coordinates
    [x2 y2] = meshgrid((-N/2 : N/2-1) / (N*d1)*wvl*Dz);
    % evaluate the Fresnel-Kirchhoff integral
    Uout = 1 / (i*wvl*Dz) ...
        .* exp(i * k/(2*Dz) * (x2.^2 + y2.^2)) ...
        .* ft2(Uin .* exp(i * k/(2*Dz) ...
        * (x1.^2 + y1.^2)), d1);
```

Listing 6.2 gives example usage of `one_step_prop` for a square aperture. Figure 6.1 shows the numerical result along with the analytic result, and it is clear

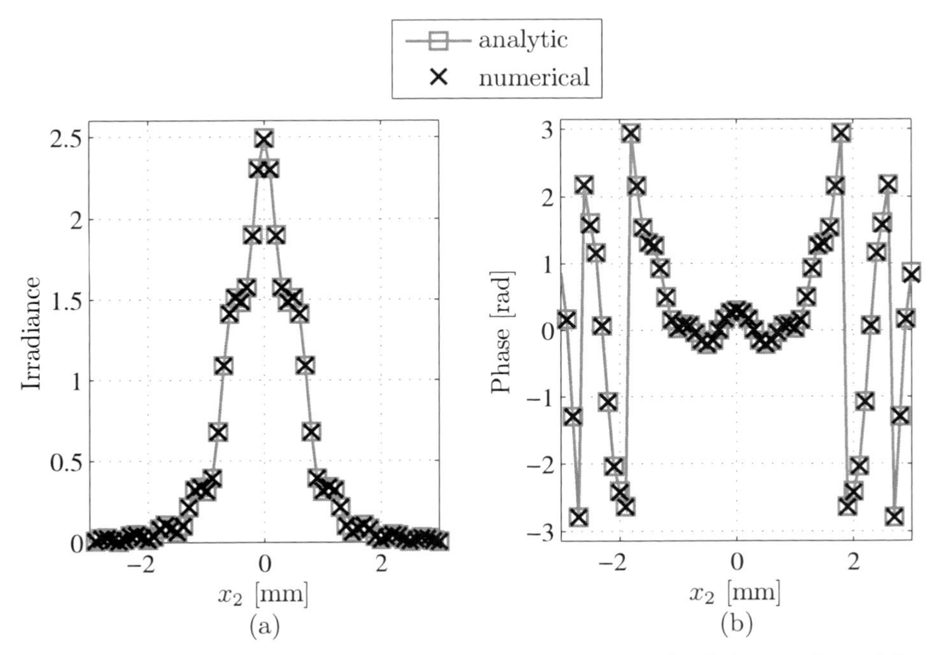

Figure 6.1 Fresnel diffraction from a square aperture, simulation and analytic: (a) observation-plane irradiance and (b) observation-plane phase.

Listing 6.2 Example of evaluating the Fresnel diffraction integral in MATLAB using a single step.

```
1  % example_square_prop_one_step.m
2
3  N = 1024;       % number of grid points per side
4  L = 1e-2;       % total size of the grid [m]
5  delta1 = L / N;   % grid spacing [m]
6  D = 2e-3;       % diameter of the aperture [m]
7  wvl = 1e-6;     % optical wavelength [m]
8  k = 2*pi / wvl;
9  Dz = 1;         % propagation distance [m]
10
11 [x1 y1] = meshgrid((-N/2 : N/2-1) * delta1);
12 ap = rect(x1/D) .* rect(y1/D);
13 [x2 y2 Uout] = one_step_prop(ap, wvl, delta1, Dz);
14
15 % analytic result for y2=0 slice
16 Uout_an ...
17     = fresnel_prop_square_ap(x2(N/2+1,:), 0, D, wvl, Dz);
```

that the comparison is very close.

Obviously, we have no control over spacing in the final grid without changing the geometry because Eq. (6.16) gives a fixed grid spacing in the observation plane. What if we have an application where the fixed value of δ_2 does not sample the observation-plane field adequately? We could obtain finer sampling in the observation plane by increasing N. Typically, we would prefer not to increase N due to the longer execution time of the simulation, though.

6.3.2 Two-step propagation

To choose the observation-plane grid spacing, we must introduce a new scaling parameter $m = \delta_2/\delta_1$. For one-step propagation [compute $U(x_2, y_2)$ directly from $U(x_1, y_1)$], there is little freedom to choose m as indicated in Eq. (6.16). Typically, λ and Δz are fixed for a given problem, so N and δ_1 must be adjusted to select a desired value of m. There must be a trade-off between the source and observation grids. A finer source grid produces a coarser observation grid and vice-versa. We could adjust N to help, but there is a practical limit to the number of grid points that can be simulated, and increasing N increases the simulation's execution time, which is typically not desirable.

Coy[35] and Rydberg and Bengtsson[36] presented a method that has more flexibility in selecting the grids. In this method, $U(x_1, y_1)$ propagates from the source plane at z_1 to an intermediate plane located at z_{1a} and then propagates to the observation plane at z_2, so that we can choose z_{1a} such that m (equivalently δ_2) has

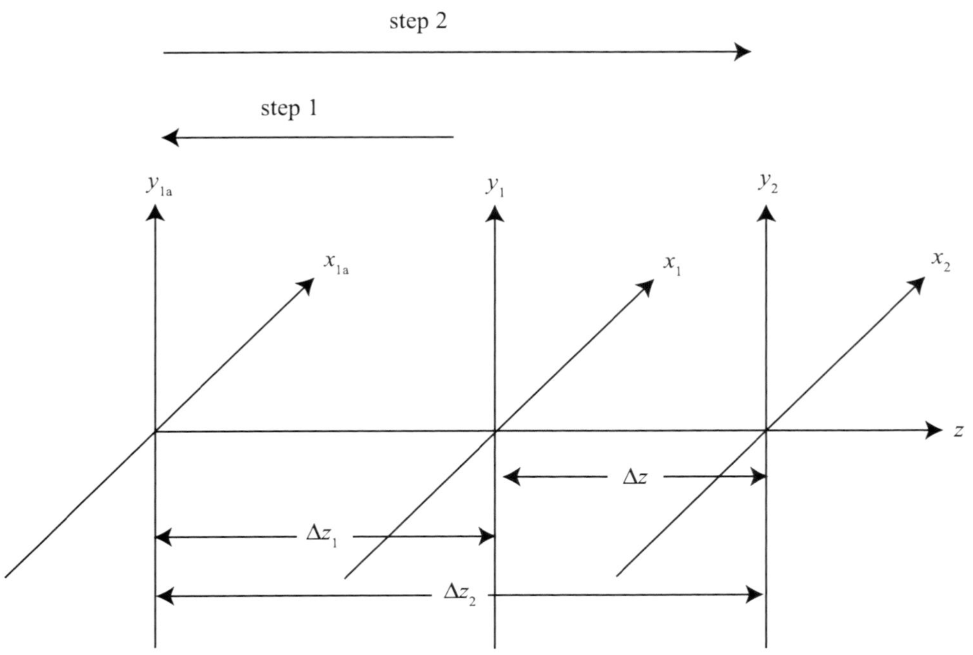

Figure 6.2 Two-step propagation geometry in which the intermediate plane is not between the source and observation planes.

the desired value. The following development follows Rydberg and Bengtsson's algorithm description with Coy's analysis of the grid spacings.

This is called two-step propagation as specified below. To keep the notation clear, the following definitions are still used: the source plane is at $z = z_1$ ($\mathbf{r}_1$ coordinates), and the observation plane is at $z = z_2$ ($\mathbf{r}_2$ coordinates) with $\Delta z = z_2 - z_1$ and scaling parameter of $m = \delta_2/\delta_1$. We define the intermediate plane at $z = z_{1a}$ [$\mathbf{r}_{1a} = (x_{1a}, y_{1a})$ coordinates] such that the distance of the first propagation is $\Delta z_1 = z_{1a} - z_1$ and the distance of the second is $\Delta z_2 = z_2 - z_{1a}$. As discussed below, there are two possible intermediate planes that yield a given scaling parameter after the two-step propagation. These two different geometries are shown in Figs. 6.2 and 6.3. In one case, the intermediate plane is far from the source and observation planes. In the other, the intermediate plane is between the source and observation planes.

In operator notation, two steps of Fresnel-integral propagation are given by

$$U(\mathbf{r}_2) = \mathcal{R}\left[\Delta z_2, \mathbf{r}_{1a}, \mathbf{r}_2\right] \mathcal{R}\left[\Delta z_1, \mathbf{r}_1, \mathbf{r}_{1a}\right] \{U(\mathbf{r}_1)\} \tag{6.17}$$

$$= \mathcal{Q}\left[\frac{1}{\Delta z_2}, \mathbf{r}\right] \mathcal{V}\left[\frac{1}{\lambda \Delta z_2}\right] \mathcal{F}\left[\mathbf{r}_2, \mathbf{f}_{1a}\right] \mathcal{Q}\left[\frac{1}{\Delta z_2}, \mathbf{r}_{1a}\right] \tag{6.18}$$

$$\times \mathcal{Q}\left[\frac{1}{\Delta z_1}, \mathbf{r}_{1a}\right] \mathcal{V}\left[\frac{1}{\lambda \Delta z_1}\right] \mathcal{F}\left[\mathbf{r}_1, \mathbf{f}_1\right] \mathcal{Q}\left[\frac{1}{\Delta z_1}, \mathbf{r}_1\right] \{U(\mathbf{r}_1)\}.$$

If we examine the spacings δ_{1a} in the intermediate plane and δ_2 in the observation

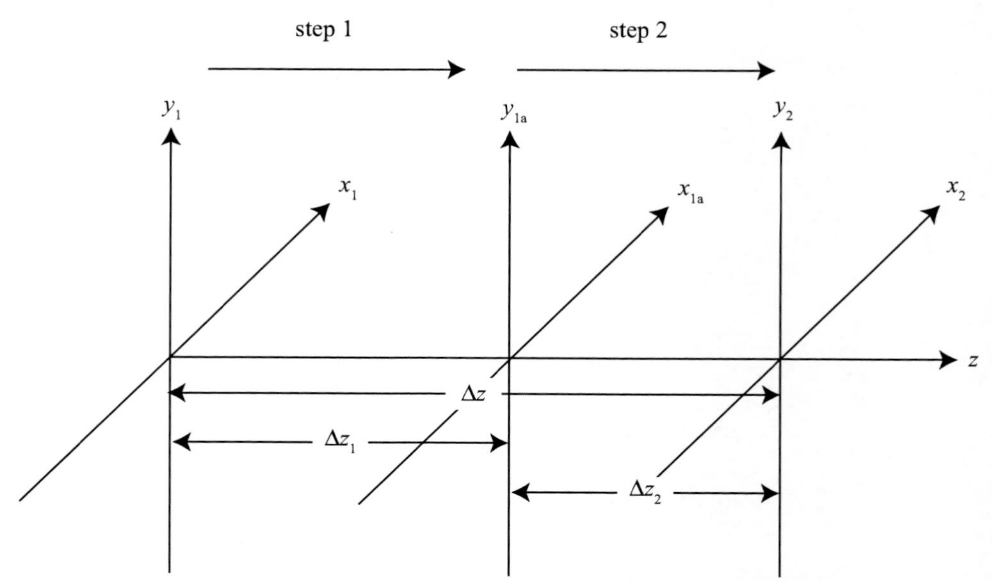

Figure 6.3 Two-step propagation geometry in which the intermediate plane is between the source and observation planes.

plane, we find

$$\delta_{1a} = \frac{\lambda\,|\Delta z_1|}{N\delta_1} \qquad \text{with} \qquad \Delta z_1 = z_{1a} - z_1 \tag{6.19}$$

$$\delta_2 = \frac{\lambda\,|\Delta z_2|}{N\delta_{1a}} \tag{6.20}$$

$$= \frac{\lambda\,|\Delta z_2|}{N\left(\frac{\lambda|\Delta z_1|}{N\delta_1}\right)} \tag{6.21}$$

$$= \left|\frac{\Delta z_2}{\Delta z_1}\right|\delta_1 \tag{6.22}$$

$$= m\delta_1, \tag{6.23}$$

which is expected given the definition of scaling parameter $m = \delta_2/\delta_1$.

Thus, a choice of m (which directly sets the sizes of the grids) defines the location of the intermediate plane, i.e., from above

$$m = \left|\frac{z_2 - z_{1a}}{z_{1a} - z_1}\right| = \left|\frac{\Delta z_2}{\Delta z_1}\right|, \tag{6.24}$$

which has solutions for the choice of z_{1a} (constrained such that $\Delta z_1 + \Delta z_2 = \Delta z$) given by

$$\Delta z_1 = z_{1a} - z_1 = \Delta z\left(\frac{1}{1 \pm m}\right) \tag{6.25}$$

Table 6.2 Examples of scaling parameter values for two-step Fresnel integral computation.

m	$\Delta z_1^+/\Delta z$	$\Delta z_2^+/\Delta z$	$\Delta z_1^-/\Delta z$	$\Delta z_2^-/\Delta z$
	$\frac{1}{(1+m)}$	$\frac{m}{(1+m)}$	$\frac{1}{(1-m)}$	$\frac{-m}{(1-m)}$
2	1/3	2/3	-1	2
1	1/2	1/2	$\pm\infty$	$\mp\infty$
1/2	2/3	1/3	2	-1

$$z_{1a} = z_1 + \Delta z \left(\frac{1}{1 \pm m} \right) \tag{6.26}$$

$$\Delta z_2 = z_2 - z_{1a} = \Delta z \left(\frac{\pm m}{1 \pm m} \right) \tag{6.27}$$

$$z_{1a} = z_2 - \Delta z \left(\frac{\pm m}{1 \pm m} \right) \tag{6.28}$$

$$z_{1a} = z_2 + \Delta z \left(\frac{\mp m}{1 \pm m} \right). \tag{6.29}$$

This has a very simple proof:

$$\left| \frac{\Delta z_2}{\Delta z_1} \right| = \left| \frac{\Delta z \left(\frac{\pm m}{1 \pm m} \right)}{\Delta z \left(\frac{1}{1 \pm m} \right)} \right| = |\pm m| = m. \tag{6.30}$$

Table 6.2 gives some example values of m with the corresponding intermediate plane locations. The Δz_1^- and Δz_2^- columns correspond to Fig. 6.2, and the Δz_1^+ and Δz_2^+ columns correspond to Fig. 6.3. Note that for unit scaling parameter, the intermediate plane is either located halfway between the source and observation planes or infinitely far away.

Listing 6.3 gives the MATLAB function `two_step_prop` that numerically evaluates Eq. (6.18). Listing 6.4 shows example usage by simply repeating the previous MATLAB example but with the two-step propagation algorithm. Figure 6.4 shows the numerical and analytic results. Note that the simulation results are identical to the analytic results again.

6.4 Angular-Spectrum Propagation

This section evaluates the convolution form of the Fresnel diffraction integral given in Eq. (6.6). We can rewrite it using the convolution theorem in operator notation as

$$U(\mathbf{r}_2) = \mathcal{F}^{-1}[\mathbf{r}_2, \mathbf{f}_1] H(\mathbf{f}_1) \mathcal{F}[\mathbf{f}_1, \mathbf{r}_1] \{U(\mathbf{r}_1)\}, \tag{6.31}$$

where $H(\mathbf{f})$ is the transfer function of free-space propagation given by

$$H(\mathbf{f}_1) = e^{ik\Delta z} e^{-i\pi\lambda\Delta z\left(f_{x1}^2 + f_{y1}^2\right)}. \tag{6.32}$$

Listing 6.3 Code for evaluating the Fresnel diffraction integral in MATLAB using two-step propagation.

```
function [x2 y2 Uout] ...
    = two_step_prop(Uin, wvl, d1, d2, Dz)
% function [x2 y2 Uout] ...
%     = two_step_prop(Uin, wvl, d1, d2, Dz)

    N = size(Uin, 1);    % number of grid points
    k = 2*pi/wvl;     % optical wavevector
    % source-plane coordinates
    [x1 y1] = meshgrid((-N/2 : 1 : N/2 - 1) * d1);
    % magnification
    m = d2/d1;
    % intermediate plane
    Dz1 = Dz / (1 - m);  % propagation distance
    d1a = wvl * abs(Dz1) / (N * d1);    % coordinates
    [x1a y1a] = meshgrid((-N/2 : N/2-1) * d1a);
    % evaluate the Fresnel-Kirchhoff integral
    Uitm = 1 / (i*wvl*Dz1) ...
        .* exp(i*k/(2*Dz1) * (x1a.^2+y1a.^2)) ...
        .* ft2(Uin .* exp(i * k/(2*Dz1) ...
        * (x1.^2 + y1.^2)), d1);
    % observation plane
    Dz2 = Dz - Dz1;  % propagation distance
    % coordinates
    [x2 y2] = meshgrid((-N/2 : N/2-1) * d2);
    % evaluate the Fresnel diffraction integral
    Uout = 1 / (i*wvl*Dz2) ...
        .* exp(i*k/(2*Dz2) * (x2.^2+y2.^2)) ...
        .* ft2(Uitm .* exp(i * k/(2*Dz2) ...
        * (x1a.^2 + y1a.^2)), d1a);
```

Equation (6.31) is known as the angular-spectrum form of the Fresnel diffraction integral, and it has been discussed and applied by many authors specifically for numerical evaluation.[28,31,32,37,40–44] Section 3.1 in this book already covers discrete convolution, which could be applicable here, but we cannot simply use the `myconv2` function from Sec. 3.1 as-is. If we did, we would have no control over the grid spacing δ_2 in the observation plane. We would be stuck with $\delta_1 = \delta_2$, corresponding to $m = 1$.

To introduce the scaling parameter m, we must go back to Eq. (6.1) and rewrite

Listing 6.4 Example of evaluating the Fresnel diffraction integral in MATLAB using two-step propagation.

```
1  % example_square_prop_two_step.m
2
3  N = 1024;      % number of grid points per side
4  L = 1e-2;     % total size of the grid [m]
5  delta1 = L / N;   % grid spacing [m]
6  D = 2e-3;     % diameter of the aperture [m]
7  wvl = 1e-6;   % optical wavelength [m]
8  k = 2*pi / wvl;
9  Dz = 1;       % propagation distance [m]
10
11 [x1 y1] = meshgrid((-N/2 : N/2-1) * delta1);
12 ap = rect(x1/D) .* rect(y1/D);
13 delta2 = wvl * Dz / (N*delta1);
14 [x2 y2 Uout] = two_step_prop(ap, wvl, delta1, delta2, Dz);
15
16 % analytic result for y2=0 slice
17 Uout_an ...
18     = fresnel_prop_square_ap(x2(N/2+1,:), 0, D, wvl, Dz);
```

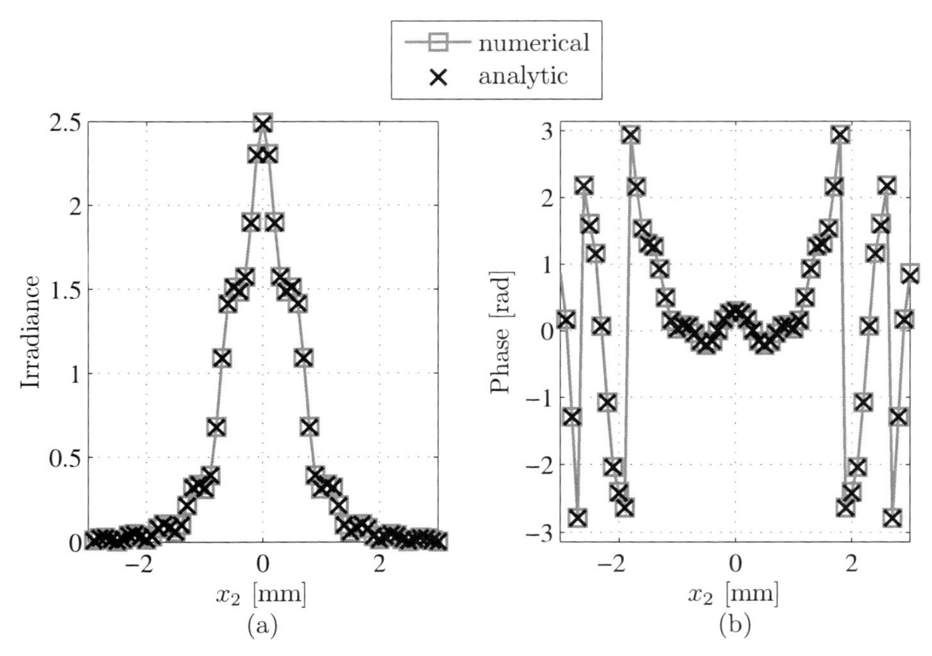

Figure 6.4 Fresnel diffraction from a square aperture, two-step simulation and analytic: (a) observation-plane irradiance and (b) observation-plane phase.

it using $\mathbf{r}_1$ and $\mathbf{r}_2$ as

$$U(\mathbf{r}_2) = \frac{1}{i\lambda\Delta z}\int_{-\infty}^{\infty} U(\mathbf{r}_1)\, e^{i\frac{k}{2\Delta z}|\mathbf{r}_2-\mathbf{r}_1|^2}\, d\mathbf{r}_1. \tag{6.33}$$

Tyler and Fried[30] and Roberts[31] are the only authors who discuss this scaling factor. Following their approach, we manipulate the exponential to introduce m:

$$|\mathbf{r}_2-\mathbf{r}_1|^2 = r_2^2 - 2\mathbf{r}_2\cdot\mathbf{r}_1 + r_1^2 \tag{6.34}$$

$$= \left(r_2^2 + \frac{r_2^2}{m} - \frac{r_2^2}{m}\right) - 2\mathbf{r}_2\cdot\mathbf{r}_1 + \left(r_1^2 + mr_1^2 - mr_1^2\right) \tag{6.35}$$

$$= \frac{r_2^2}{m} + \left(1-\frac{1}{m}\right)r_2^2 - 2\mathbf{r}_2\cdot\mathbf{r}_1 + \left[mr_1^2 + (1-m)\,r_1^2\right] \tag{6.36}$$

$$= m\left[\left(\frac{r_2}{m}\right)^2 - 2\left(\frac{\mathbf{r}_2}{m}\right)\cdot\mathbf{r}_1 + r_1^2\right] + \left(1-\frac{1}{m}\right)r_2^2 + (1-m)\,r_1^2 \tag{6.37}$$

$$= m\left|\frac{\mathbf{r}_2}{m} - \mathbf{r}_1\right|^2 - \left(\frac{1-m}{m}\right)r_2^2 + (1-m)\,r_1^2. \tag{6.38}$$

Then, we can substitute it back into Eq. (6.33) to get

$$U(\mathbf{r}_2) = \frac{1}{i\lambda\Delta z}\int_{-\infty}^{\infty} U(\mathbf{r}_1)\, e^{i\frac{k}{2\Delta z}\left[m\left|\frac{\mathbf{r}_2}{m}-\mathbf{r}_1\right|^2 - \left(\frac{1-m}{m}\right)r_2^2 + (1-m)r_1^2\right]}\, d\mathbf{r}_1 \tag{6.39}$$

$$= \frac{e^{-i\frac{k}{2\Delta z}\left(\frac{1-m}{m}\right)r_2^2}}{i\lambda\Delta z}\int_{-\infty}^{\infty} U(\mathbf{r}_1)\, e^{i\frac{k}{2\Delta z}(1-m)r_1^2}\, e^{i\frac{km}{2\Delta z}\left|\frac{\mathbf{r}_2}{m}-\mathbf{r}_1\right|^2}\, d\mathbf{r}_1. \tag{6.40}$$

We start on the path back to obtaining a convolution integral by defining

$$U''(\mathbf{r}_1) \equiv \frac{1}{m}U(\mathbf{r}_1)\, e^{i\frac{k}{2\Delta z}(1-m)r_1^2}, \tag{6.41}$$

and substitute it into Eq. (6.40) to get

$$U(\mathbf{r}_2) = \frac{e^{-i\frac{k}{2\Delta z}\left(\frac{1-m}{m}\right)r_2^2}}{i\lambda\Delta z}\int_{-\infty}^{\infty} mU''(\mathbf{r}_1)\, e^{i\frac{km}{2\Delta z}\left|\frac{\mathbf{r}_2}{m}-\mathbf{r}_1\right|^2}\, d\mathbf{r}_1. \tag{6.42}$$

Then, defining the scaled coordinate and distance

$$\mathbf{r}_2' = \frac{\mathbf{r}_2}{m} \tag{6.43}$$

$$\Delta z' = \frac{\Delta z}{m}, \tag{6.44}$$

we obtain

$$U\left(m\mathbf{r}_2'\right) = \frac{e^{-i\frac{k}{2\Delta z'}(1-m)\left(r_2'\right)^2}}{i\lambda\Delta z'} \int_{-\infty}^{\infty} U''\left(\mathbf{r}_1\right) e^{i\frac{k}{2\Delta z'}\left|\mathbf{r}_2'-\mathbf{r}_1\right|^2} d\mathbf{r}_1. \tag{6.45}$$

Finally, this is in the form of a convolution so that

$$U\left(m\mathbf{r}_2'\right) = e^{-i\frac{k}{2\Delta z'}(1-m)\left(r_2'\right)^2} \int_{-\infty}^{\infty} U''\left(\mathbf{r}_1\right) h\left(\mathbf{r}_2' - \mathbf{r}_1\right) d\mathbf{r}_1, \tag{6.46}$$

$$\text{with } h\left(\mathbf{r}_1\right) = \frac{1}{i\lambda\Delta z'} e^{i\frac{k}{2\Delta z'}r_1^2}. \tag{6.47}$$

Once again, propagation can be treated as a linear system with a known impulse response (amplitude spread function). The FT of the impulse response is the amplitude transfer function, given by

$$\mathcal{F}\left[\mathbf{r}_1, \mathbf{f}_1\right] h\left(\mathbf{r}_1\right) = H\left(\mathbf{f}_1\right) \tag{6.48}$$

$$= e^{-i\pi\lambda\Delta z' f_1^2}. \tag{6.49}$$

At this point, we could evaluate Eq. (6.46) numerically using `myconv2`. However, using the convolution theorem and substituting back to original coordinates allows us to keep all of the details of this algorithm manifest and thereby make some simplifications in later chapters. Applying the convolution theorem leads to

$$\begin{aligned}
U'\left(m\mathbf{r}_2'\right) &= \mathcal{F}^{-1}\left[\mathbf{f}_1, \mathbf{r}_2'\right] e^{-i\pi\lambda\Delta z' f_1^2} \mathcal{F}\left[\mathbf{r}_1, \mathbf{f}_1\right] \left\{U''\left(\mathbf{r}_1\right)\right\} \\
U'\left(\mathbf{r}_2\right) &= \mathcal{F}^{-1}\left[\mathbf{f}_1, \frac{\mathbf{r}_2}{m}\right] e^{-i\pi\lambda\frac{\Delta z}{m} f_1^2} \mathcal{F}\left[\mathbf{r}_1, \mathbf{f}_1\right] \left\{U''\left(\mathbf{r}_1\right)\right\} \\
U\left(\mathbf{r}_2\right) &= e^{-i\frac{k}{2\Delta z}m(1-m)\left(\frac{\mathbf{r}_2}{m}\right)^2} \mathcal{F}^{-1}\left[\mathbf{f}_1, \frac{\mathbf{r}_2}{m}\right] e^{-i\frac{\pi\lambda\Delta z}{m} f_1^2} \\
&\quad \times \mathcal{F}\left[\mathbf{r}_1, \mathbf{f}_1\right] \left\{\frac{1}{m} U\left(\mathbf{r}_1\right) e^{i\frac{k}{2\Delta z}(1-m)r_1^2}\right\} \\
&= e^{-i\frac{k}{2\Delta z}\frac{1-m}{m}r_2^2} \mathcal{F}^{-1}\left[\mathbf{f}_1, \frac{\mathbf{r}_2}{m}\right] e^{-i\frac{\pi\lambda\Delta z}{m} f_1^2} \\
&\quad \times \mathcal{F}\left[\mathbf{r}_1, \mathbf{f}_1\right] \left\{\frac{1}{m} U\left(\mathbf{r}_1\right) e^{i\frac{k}{2\Delta z}(1-m)r_1^2}\right\} \\
&= \mathcal{Q}\left[\frac{m-1}{m\Delta z}, \mathbf{r}_2\right] \mathcal{F}^{-1}\left[\mathbf{f}_1, \frac{\mathbf{r}_2}{m}\right] \mathcal{Q}_2\left[-\frac{\Delta z}{m}, \mathbf{f}_1\right] \\
&\quad \times \mathcal{F}\left[\mathbf{r}_1, \mathbf{f}_1\right] \mathcal{Q}\left[\frac{1-m}{\Delta z}, \mathbf{r}_1\right] \frac{1}{m} \left\{U\left(\mathbf{r}_1\right)\right\}.
\end{aligned} \tag{6.50}$$

Now that we have an expression of angular-spectrum propagation in terms of operators, we can examine grid spacings δ_1 in the source plane, δ_{f1} in the spatial-frequency plane, and δ_2 in the observation plane:

$$\delta_{f1} = \frac{1}{N\delta_1} \quad \text{from } \mathcal{F}\left[\mathbf{r}_1, \mathbf{f}_1\right] \tag{6.51}$$

$$\delta_2 = \frac{m}{N\delta_{f1}} \quad \text{from } \mathcal{F}^{-1}\left[\mathbf{f}_1, \mathbf{r}_2/m\right] \tag{6.52}$$

$$= \frac{m}{N\left(\frac{1}{N\delta_1}\right)} \tag{6.53}$$

$$= m\delta_1. \tag{6.54}$$

This last equation is a consistency check. Also, we can determine two other relationships:

$$\frac{1}{1-m} = \frac{1}{1-\frac{\delta_2}{\delta_1}} = \frac{\delta_1}{\delta_1 - \delta_2} \tag{6.55}$$

$$\frac{m}{1-m} = \frac{\frac{\delta_2}{\delta_1}}{1-\frac{\delta_2}{\delta_1}} = \frac{\delta_2}{\delta_1 - \delta_2}. \tag{6.56}$$

These relationships are used later in Sec. 8.2.

Another solution for the angular-spectrum formulation can be found. Let us start at Eq. (6.34) to manipulate $|\mathbf{r}_2 - \mathbf{r}_1|^2$ a little differently:

$$|\mathbf{r}_2 - \mathbf{r}_1|^2 = r_2^2 - 2\mathbf{r}_2 \cdot \mathbf{r}_1 + r_1^2 \tag{6.57}$$

$$= \left(r_2^2 + \frac{r_2^2}{m} - \frac{r_2^2}{m}\right) - 2\mathbf{r}_2 \cdot \mathbf{r}_1 + \left(r_1^2 + mr_1^2 - mr_1^2\right) \tag{6.58}$$

$$= -\frac{r_2^2}{m} + \left(1 + \frac{1}{m}\right) r_2^2 - 2\mathbf{r}_2 \cdot \mathbf{r}_1 - mr_1^2 + (1+m)\, r_1^2 \tag{6.59}$$

$$= -\frac{r_2^2}{m} - 2\mathbf{r}_2 \cdot \mathbf{r}_1 - mr_1^2 + \left(1 + \frac{1}{m}\right) r_2^2 + (1+m)\, r_1^2 \tag{6.60}$$

$$= -m\left(\left|\frac{\mathbf{r}_2}{m}\right|^2 + 2\left(\frac{\mathbf{r}_2}{m}\right) \cdot \mathbf{r}_1 + r_1^2\right) + \left(1 + \frac{1}{m}\right) r_2^2 + (1+m)\, r_1^2 \tag{6.61}$$

$$= -m\left|\frac{\mathbf{r}_2}{m} + \mathbf{r}_1\right|^2 + \left(\frac{1+m}{m}\right) r_2^2 + (1+m)\, r_1^2. \tag{6.62}$$

With a substitution of $m' = -m$,

$$= m'\left|\frac{\mathbf{r}_2}{-m'} + \mathbf{r}_1\right|^2 + \left(\frac{1-m'}{-m'}\right) r_2^2 + \left(1 - m'\right) r_1^2 \tag{6.63}$$

$$= m'\left|\frac{\mathbf{r}_2}{m'} - \mathbf{r}_1\right|^2 - \left(\frac{1-m'}{m'}\right) r_2^2 + \left(1 - m'\right) r_1^2, \tag{6.64}$$

it is obvious that this is identical to Eq. (6.38) with the use of m' rather than m.

Now with the realization that $\pm m$ may be used in the angular-spectrum form of diffraction, there are two possible equations:

$$U(\mathbf{r}_2) = \mathcal{Q}\left[\frac{m-1}{m\Delta z}, \mathbf{r}_2\right] \mathcal{F}^{-1}\left[\mathbf{f}_1, \frac{\mathbf{r}_2}{m}\right] \mathcal{Q}_2\left[-\frac{\Delta z}{m}, \mathbf{f}_1\right]$$

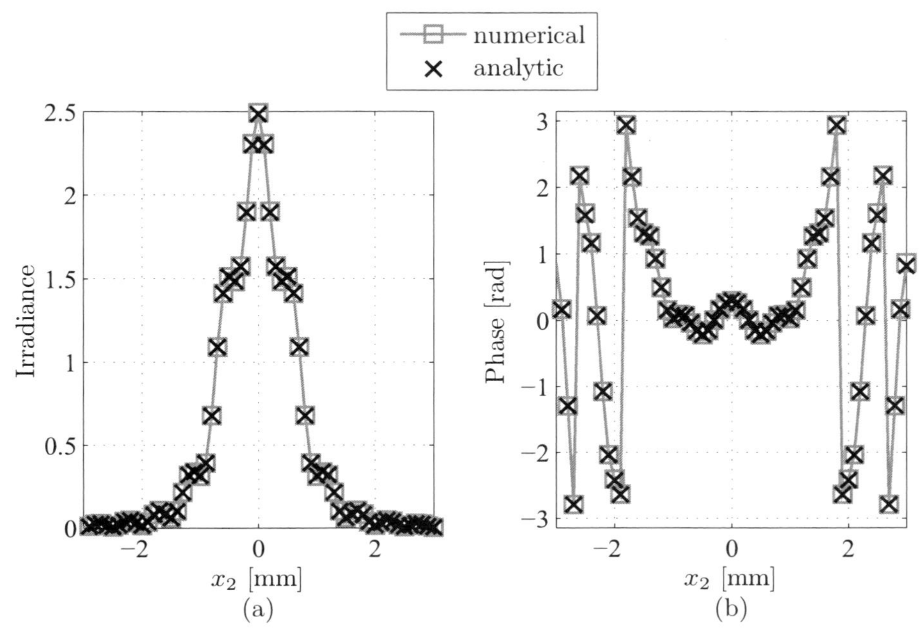

Figure 6.5 Fresnel diffraction from a square aperture, angular-spectrum simulation and analytic: (a) observation-plane irradiance and (b) observation-plane phase.

$$\times \mathcal{F}\left[\mathbf{r}_1, \mathbf{f}_1\right] \mathcal{Q}\left[\frac{1-m}{\Delta z}, \mathbf{r}_1\right] \frac{1}{m}\left\{U\left(\mathbf{r}_1\right)\right\} \tag{6.65}$$

$$= \mathcal{Q}\left[-\frac{m-1}{m\Delta z}, \mathbf{r}_2\right] \mathcal{F}^{-1}\left[\mathbf{f}_1, \frac{\mathbf{r}_2}{m}\right] \mathcal{Q}_2\left[\frac{\Delta z}{m}, \mathbf{f}_1\right]$$

$$\times \mathcal{F}\left[\mathbf{r}_1, \mathbf{f}_1\right] \mathcal{Q}\left[-\frac{1-m}{\Delta z}, \mathbf{r}_1\right] \left(\frac{-1}{m}\right)\left\{U\left(\mathbf{r}_1\right)\right\} \tag{6.66}$$

This can be written more compactly as

$$U(\mathbf{r}) = \mathcal{Q}\left[\frac{m \pm 1}{m\Delta z}, \mathbf{r}_2\right] \mathcal{F}^{-1}\left[\mathbf{f}_1, \mp\frac{\mathbf{r}_2}{m}\right] \mathcal{Q}_2\left[\pm\frac{\Delta z}{m}, \mathbf{f}_1\right]$$

$$\times \mathcal{F}\left[\mathbf{r}_1, \mathbf{f}_1\right] \mathcal{Q}\left[\frac{1 \pm m}{\Delta z}, \mathbf{r}_1\right] \left(\mp\frac{1}{m}\right)\left\{U\left(\mathbf{r}_1\right)\right\}, \tag{6.67}$$

where the top sign corresponds to Eq. (6.66), and the bottom sign corresponds to Eq. (6.65).

Listing 6.5 gives the MATLAB function `ang_spec_prop` that numerically evaluates Eq. (6.65). Figure 6.5 shows the results of repeating the previous MATLAB examples using angular-spectrum propagation. The code that produced Fig. 6.5 is not shown here because it is identical to Listing 6.4 except for line 14, which calls the function `ang_spec_prop` given in Listing 6.5. Note that the numerical results are identical to the analytic results again.

Listing 6.5 Example of evaluating the Fresnel diffraction integral in MATLAB using the angular-spectrum method.

```
function [x2 y2 Uout] ...
    = ang_spec_prop(Uin, wvl, d1, d2, Dz)
% function [x2 y2 Uout] ...
%     = ang_spec_prop(Uin, wvl, d1, d2, Dz)

    N = size(Uin,1);   % assume square grid
    k = 2*pi/wvl;      % optical wavevector
    % source-plane coordinates
    [x1 y1] = meshgrid((-N/2 : 1 : N/2 - 1) * d1);
    r1sq = x1.^2 + y1.^2;
    % spatial frequencies (of source plane)
    df1 = 1 / (N*d1);
    [fX fY] = meshgrid((-N/2 : 1 : N/2 - 1) * df1);
    fsq = fX.^2 + fY.^2;
    % scaling parameter
    m = d2/d1;
    % observation-plane coordinates
    [x2 y2] = meshgrid((-N/2 : 1 : N/2 - 1) * d2);
    r2sq = x2.^2 + y2.^2;
    % quadratic phase factors
    Q1 = exp(i*k/2*(1-m)/Dz*r1sq);
    Q2 = exp(-i*pi^2*2*Dz/m/k*fsq);
    Q3 = exp(i*k/2*(m-1)/(m*Dz)*r2sq);
    % compute the propagated field
    Uout = Q3.* ift2(Q2 .* ft2(Q1 .* Uin / m, d1), df1);
```

6.5 Simple Optical Systems

Most of the wave propagation simulations in this book are through either vacuum or weakly refractive media like atmospheric turbulence. Moreover, the whole formalism presented up to this point can be extended to simple refractive and reflective optical systems. The effect of such simple systems is described through geometric optics by the use of paraxial ray matrices.[45]

Ray matrices describe how a refractive element transforms the location and direction of paraxial rays. In this framework, rays are represented by their ray height y_1 (distance from the optical axis at a certain z location), ray slope y_1', and the refractive index n_1 of the medium that contains the ray. Usually rays are confined to the marginal $(y - z)$ plane. As a ray passes through a simple optical system, the system's effect on the ray is represented by a system of two coupled linear equations:

$$y_2 = A\,y_1 + B\,n_1\,y_1' \tag{6.68}$$

$$n_2\, y_2' = C\, y_1 + D\, n_1\, y_1', \tag{6.69}$$

where y_2, y_2', and n_2 are the ray height, slope, and refractive index, respectively, after the optical system. This way, the system is characterized by the values of A, B, C, and D. This can be written in matrix-vector notation as

$$\begin{pmatrix} y_2 \\ n_2\, y_2' \end{pmatrix} = \begin{pmatrix} A & B \\ C & D \end{pmatrix} \begin{pmatrix} y_1 \\ n_1\, y_1' \end{pmatrix}. \tag{6.70}$$

Note that ray matrices are always written so that $AD - BC = 1$.

There are two elementary ray matrices: that for ray transfer and that for refraction. Ray transfer simply refers to pure propagation, and refraction means that the ray encounters a surface that forms the interface between two materials of unlike refractive index. With ray transfer, the ray slope remains the same, and the ray height increases according to the ray slope and propagation distance so that[45]

$$\begin{pmatrix} y_2 \\ n_2\, y_2' \end{pmatrix} = \begin{pmatrix} 1 & \Delta z/n_1 \\ 0 & 1 \end{pmatrix} \begin{pmatrix} y_1 \\ n_1\, y_1' \end{pmatrix}. \tag{6.71}$$

With refraction, the ray height remains the same, but the ray slope changes according to the paraxial version of Snell's law so that

$$\begin{pmatrix} y_2 \\ n_2\, y_2' \end{pmatrix} = \begin{pmatrix} 1 & 0 \\ \frac{n_1 - n_2}{R} & 1 \end{pmatrix} \begin{pmatrix} y_1 \\ n_1\, y_1' \end{pmatrix}, \tag{6.72}$$

where R is the surface's radius of curvature.[45]

Without regard to vignetting, optical systems can be modeled as the successive application of ray transfer and refraction matrices written right-to-left. For example, a light ray passing from air just before the front face of a singlet lens of index n to just after the back end of the lens encounters refraction at the first surface, transfer through the lens, and then refraction at the back interface, represented by the system matrix

$$\mathsf{S} = \begin{pmatrix} 1 & 0 \\ \frac{n-1}{R_2} & 1 \end{pmatrix} \begin{pmatrix} 1 & \Delta z/n \\ 0 & 1 \end{pmatrix} \begin{pmatrix} 1 & 0 \\ \frac{1-n}{R_1} & 1 \end{pmatrix}. \tag{6.73}$$

In this equation, R_1 and R_2 are the radii of curvature of the two lens faces. If the lens is thin enough that $\Delta z \approx 0$, then the lens matrix simplifies to

$$\mathsf{S} = \begin{pmatrix} 1 & 0 \\ (1-n)\left(\frac{1}{R_1} - \frac{1}{R_2}\right) & 1 \end{pmatrix}. \tag{6.74}$$

Now, the lensmaker's equation gives the focal length f_l of a lens in terms its radii and index according to

$$\frac{1}{f_l} = (n-1)\left(\frac{1}{R_1} - \frac{1}{R_2}\right). \tag{6.75}$$

When this is used, the lens matrix becomes

$$\mathsf{S} = \begin{pmatrix} 1 & 0 \\ -1/f_l & 1 \end{pmatrix}. \tag{6.76}$$

Diffraction calculations account for simple optical systems through the generalized Huygens-Fresnel integral given by[15,34,46–48]

$$U(x_2, y_2) = \frac{e^{ikz}}{i\lambda B} \int_{-\infty}^{\infty} \int_{-\infty}^{\infty} U(x_1, y_1)\, e^{i\frac{k}{2B}\left(Dr_2^2 - 2\mathbf{r}_1\cdot\mathbf{r}_2 + Ar_1^2\right)}\, dx_1 dy_1. \tag{6.77}$$

Note that this is valid only for optical systems possessing azimuthal symmetry, such as circular lenses with spherical radii of curvature on each face. Eq. (6.77) can be easily generalized for non-symmetric systems like square apertures, cylindrical lenses, and toroidal lenses.[47] This integral is closely related to the fractional Fourier transform.[49] Numerical implementations have been implemented numerically by several authors.[34,50–52]

There are two particularly interesting cases to note here. For pure ray transfer, $A = D = 1$, $C = 0$, and $B = \Delta z$ so that Eq. (6.77) reduces to the free-space Fresnel diffraction integral in Eq. (6.1), as it should. When the light propagates from the front face of a spherical lens to its back focal plane, $A = 0$, $B = f_l$, $C = -f_l^{-1}$, and $D = 1$ so that Eq. (6.77) reduces to a scaled FT, much like in Eq. (4.9).

The generalized Huygens-Fresnel integral is more complicated than the Fresnel diffraction integral, and at first glance it may not appear like a convolution integral. However, Lambert and Fraser showed that simple substitutions can transform it into a convolution so that the computational methods discussed in the previous sections of this chapter may be applied.[47] Following their method, we substitute

$$\alpha = \frac{A}{\lambda B} \text{ and } \beta = \frac{AC}{\lambda} \tag{6.78}$$

and recall that $AD - BC = 1$ to obtain[47]

$$U(A\mathbf{r}_2) = \frac{1}{i\lambda B} e^{i\pi\beta r_2^2} \int_{-\infty}^{\infty} U(\mathbf{r}_1)\, e^{i\pi\alpha|\mathbf{r}_2 - \mathbf{r}_1|^2}\, d\mathbf{r}_1. \tag{6.79}$$

This is clearly a convolution, and we can write it explicitly as

$$U(A\mathbf{r}_2) = \frac{1}{i\lambda B} e^{i\pi\beta r_2^2} \left[U(\mathbf{r}_1) \otimes e^{i\pi\alpha r_1^2}\right]. \tag{6.80}$$

Further, we can see that the transfer function for the optical system is

$$H(\mathbf{f}) = \frac{i}{\alpha} e^{-i\frac{\pi}{\alpha}\left(f_x^2 + f_y^2\right)}. \tag{6.81}$$

Listing 6.6 Code for evaluating the Fresnel diffraction integral in MATLAB using the angular-spectrum method with an ABCD ray matrix.

```
1  function [x2 y2 Uout] ...
2      = ang_spec_propABCD(Uin, wvl, d1, d2, ABCD)
3  % function [x2 y2 Uout] ...
4  %     = ang_spec_propABCD(Uin, wwl, d1, d2, ABCD)
5
6      N = size(Uin,1);    % assume square grid
7      k = 2*pi/wvl;      % optical wavevector
8      % source-plane coordinates
9      [x1 y1] = meshgrid((-N/2 : 1 : N/2 - 1) * d1);
10     r1sq = x1.^2 + y1.^2;
11     % spatial frequencies (of source plane)
12     df1 = 1 / (N*d1);
13     [fX fY] = meshgrid((-N/2 : 1 : N/2 - 1) * df1);
14     fsq = fX.^2 + fY.^2;
15     % scaling parameter
16     m = d2/d1;
17     % observation-plane coordinates
18     [x2 y2] = meshgrid((-N/2 : 1 : N/2 - 1) * d2);
19     r2sq = x2.^2 + y2.^2;
20     % optical system matrix
21     A = ABCD(1,1); B = ABCD(1,2); D = ABCD(2,2);
22     % quadratic phase factors
23     Q1 = exp(i*pi/(wvl*B)*(A-m)*r1sq);
24     Q2 = exp(-i*pi*wvl*B/m*fsq);
25     Q3 = exp(i*pi/(wvl*B)*(D-1/m)*r2sq);
26     % compute the propagated field
27     Uout = Q3.* ift2(Q2 .* ft2(Q1 .* Uin / m, d1), df1);
```

Recall that this algorithm does not account for vignetting of the rays due to finite-extent apertures in the optical system. The most straightforward way to handle this is to simulate propagation from aperture to aperture, setting the vignetted portions to zero at each aperture. However, the reader is directed to Coy for a more detailed and efficient method of accounting for vignetting in simulations.[35]

Listing 6.6 gives the MATLAB function `ang_spec_propABCD` that evaluates Eq. (6.79). Figure 6.6 shows the results of repeating the previous MATLAB examples using angular-spectrum propagation, using an ABCD ray matrix to represent the free space. The code that produced Fig. 6.6 is given in Listing 6.7. Note that the numerical results are identical to the analytic results again.

Listing 6.7 Example of propagating light from a square aperture using the ABCD ray-matrix simulation method.

```
1  % example_square_prop_ang_specABCD.m
2
3  N = 1024;      % number of grid points per side
4  L = 1e-2;     % total size of the grid [m]
5  delta1 = L / N;  % grid spacing [m]
6  D = 2e-3;     % diameter of the aperture [m]
7  wvl = 1e-6;  % optical wavelength [m]
8  k = 2*pi / wvl;
9  Dz = 1;       % propagation distance [m]
10 f = inf;      % source field radius of curvature [m]
11
12 [x1 y1] = meshgrid((-N/2 : N/2-1) * delta1);
13 ap = rect(x1/D) .* rect(y1/D);
14 delta2 = wvl * Dz / (N*delta1);
15
16 ABCD = [1 Dz; 0 1] * [1 0 ; -1/f 1];
17 [x2 y2 Uout] ...
18     = ang_spec_propABCD(ap, wvl, delta1, delta2, ABCD);
```

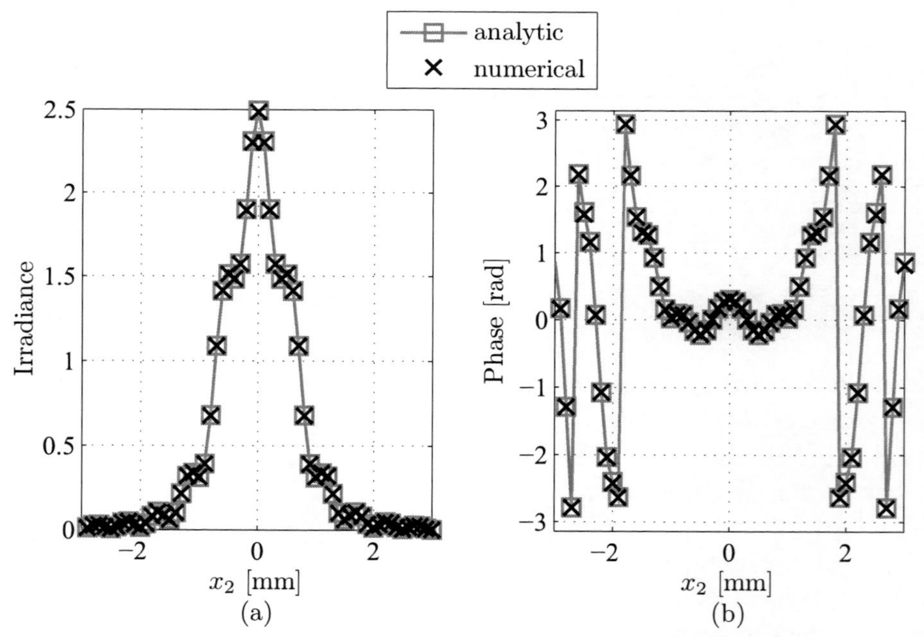

Figure 6.6 Observation-plane field resulting from square-aperture source with a diverging spherical wavefront. This simulation used the ABCD ray-matrix method of propagation.

Listing 6.8 Example of propagating a sinc model point source in MATLAB using the angular-spectrum method.

```
1  % example_pt_source.m
2
3  D = 8e-3;    % diameter of the observation aperture [m]
4  wvl = 1e-6;  % optical wavelength [m]
5  k = 2*pi / wvl; % optical wavenumber [rad/m]
6  Dz = 1;      % propagation distance [m]
7  arg = D/(wvl*Dz);
8  delta1 = 1/(10*arg); % source-plane grid spacing [m]
9  delta2 = D/100; % observation-plane grid spacing [m]
10 N = 1024;          % number of grid points
11 % source-plane coordinates
12 [x1 y1] = meshgrid((-N/2 : N/2-1) * delta1);
13 [theta1 r1] = cart2pol(x1, y1);
14 A = wvl * Dz;    % sets field amplitude to 1 in obs plane
15 pt = A * exp(-i*k/(2*Dz) * r1.^2) * arg^2 ...
16     .* sinc(arg*x1) .* sinc(arg*y1);
17 [x2 y2 Uout] = ang_spec_prop(pt, wvl, delta1, delta2, Dz);
```

6.6 Point Sources

Point sources are especially challenging to model. Recall from Ch. 1 that a true point source $U_{pt}(\mathbf{r}_1)$ is represented by a Dirac delta function via

$$U_{pt}(\mathbf{r}_1) = \delta(\mathbf{r}_1 - \mathbf{r}_c), \tag{6.82}$$

where $\mathbf{r}_c = (x_c, y_c)$ is the location of the point source in the $x_1 - y_1$ plane. The field $U_{pt}(\mathbf{r}_1)$ has a Fourier spectrum that is constant across all spatial frequencies. This means that it has infinite spatial bandwidth, which is unusual because most optical sources are spatially bandlimited. The infinite spatial bandwidth is a problem for the discretely sampled and finite-sized grid that we must use in computer simulations. If a propagation grid has spacing δ_1 in the source plane, then the highest spatial frequency represented on that grid without aliasing is $1/(2\delta_1)$. Therefore, a bandlimited version of a point source must suffice. The point source in the simulation must have a finite spatial extent.

Various point-source models have been used in the literature. To simulate propagation though turbulence, Martin and Flatté[44] and Coles[32] used a narrow Gaussian function with a quadratic phase. Martin and Flatté's model point source is given by

$$\exp\left(-\frac{r^2}{2\sigma^2}\right)\exp\left(-i\frac{r^2}{2x_0^2}\right). \tag{6.83}$$

The parameters σ and x_0 were equal to the grid spacing. This is similar to the example from Sec. 2.5.3. With use of absorbing boundaries in the simulation (discussed

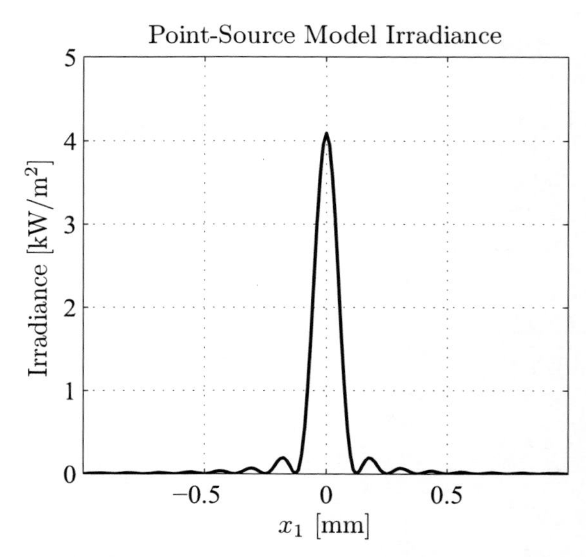

Figure 6.7 Irradiance of a sinc model of a point source (source plane).

in Sec. 8.1), this model produced an observation-plane field that was approximately flat across the central one-third of their propagation grid and tapered to zero toward the edge. Later, Flatté *et al.*[53] used a model point-source field given by

$$\exp\left(-\frac{r^2}{2\sigma^2}\right)\cos^2\left(\frac{r^2}{2\rho^2}\right), \tag{6.84}$$

where σ and ρ are nearly equal to the grid spacing. This model also produced a field that was approximately flat across the central one-third of their observation-plane grid and tapered to zero toward the edge.

Here, we take a different approach and seek a good model by analytically computing the desired observation-plane field. If we observe the field in the $x_2 - y_2$ plane a distance Δz away from the source, we can easily evaluate Eq. (6.1), (6.5), or (6.18) to obtain the field, given by

$$U(\mathbf{r}_2) = \frac{e^{ik\Delta z}}{i\lambda\Delta z} e^{i\frac{k}{2\Delta z}|\mathbf{r}_2-\mathbf{r}_c|^2}. \tag{6.85}$$

This result is the paraxial approximation to a spherical wave. It has constant amplitude across the $x_2 - y_2$ plane and a parabolic phase.

Our goal is to obtain good agreement between the simulation and potential experiments. Any camera or wavefront sensor that we might use occupies only a finite region of the $x_2 - y_2$ plane. Therefore, our source model is valid if our simulation obtains good agreement over the detector area. Then, let us work with a field $\widetilde{U}(\mathbf{r}_2)$ that has finite spatial extent, given by

$$\widetilde{U}(\mathbf{r}_2) = \frac{e^{ik\Delta z}}{i\lambda\Delta z} W(\mathbf{r}_2 - \mathbf{r}_c)\, e^{i\frac{k}{2\Delta z}|\mathbf{r}_2-\mathbf{r}_c|^2}, \tag{6.86}$$

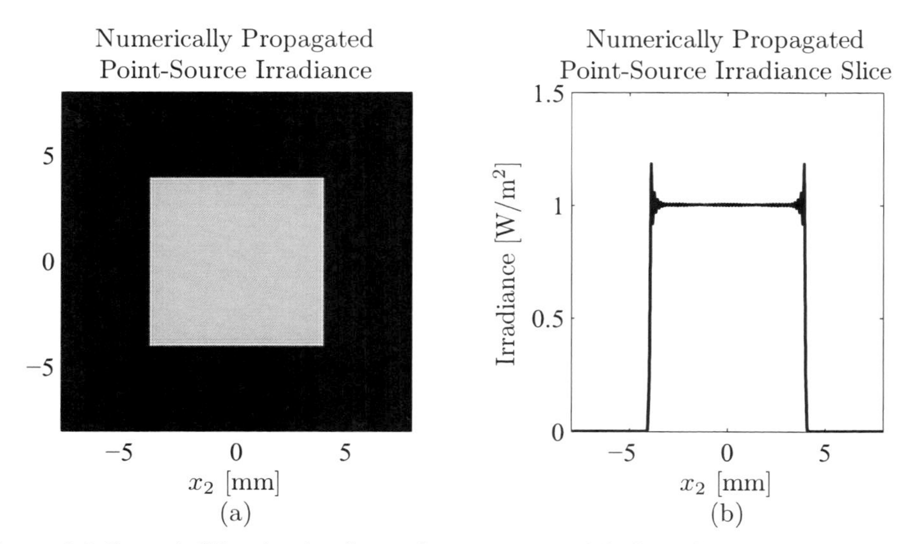

Figure 6.8 Fresnel diffraction irradiance from a sinc model of a point source (observation plane).

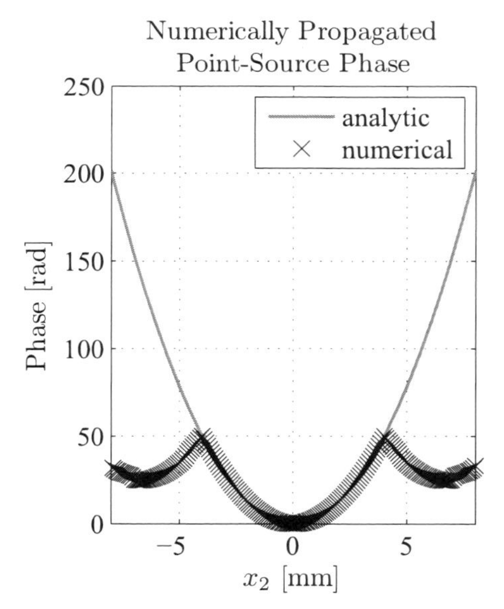

Figure 6.9 Fresnel diffraction phase from a sinc model of a point source (observation plane).

where $W(\mathbf{r}_2)$ is a "window" function that is nonzero over only a finite region of space. The extent of $W(\mathbf{r}_2)$ must be at least as large as the detector, but smaller than the propagation grid. For example, it might be a two-dimensional rect or circ function.

Let us represent our point-source model by $\widetilde{U}_{pt}(\mathbf{r}_1)$, substitute it into the Fresnel diffraction integral, and set the result equal to $\widetilde{U}(\mathbf{r}_2)$:

$$\widetilde{U}(\mathbf{r}_2) = \frac{e^{ik\Delta z}}{i\lambda\Delta z} e^{i\frac{k}{2\Delta z}r_2^2} \int_{-\infty}^{\infty} \widetilde{U}_{pt}(\mathbf{r}_1)\, e^{i\frac{k}{2\Delta z}r_1^2} e^{-i\frac{2\pi}{\lambda\Delta z}\mathbf{r}_1\cdot\mathbf{r}_2}\, d\mathbf{r}_1$$

$$= \frac{e^{ik\Delta z}}{i\lambda\Delta z} e^{i\frac{k}{2\Delta z}r_2^2} \mathcal{F}\left\{\widetilde{U}_{pt}(\mathbf{r}_1)\, e^{i\frac{k}{2\Delta z}r_1^2}\right\}_{\mathbf{f}_1=\frac{\mathbf{r}_2}{\lambda\Delta z}}. \tag{6.87}$$

Then, we can solve for the point-source model given by

$$\widetilde{U}_{pt}(\mathbf{r}_1) = i\lambda\Delta z\, e^{-ik\Delta z} e^{-i\frac{k}{2\Delta z}r_1^2} \mathcal{F}^{-1}\left\{\widetilde{U}(\lambda\Delta z\mathbf{f}_1)\, e^{-i\pi\lambda\Delta z f_1^2}\right\}. \tag{6.88}$$

Now, substituting Eq. (6.86) for $\widetilde{U}(\lambda\Delta z\mathbf{f}_1)$ yields

$$\widetilde{U}_{pt}(\mathbf{r}_1) = e^{-i\frac{k}{2\Delta z}r_1^2} e^{i\frac{k}{2\Delta z}r_c^2} \mathcal{F}^{-1}\left\{W(\lambda\Delta z\mathbf{f}_1 - \mathbf{r}_c)\, e^{-i2\pi\mathbf{r}_c\cdot\mathbf{f}_1}\right\}. \tag{6.89}$$

For example, if a square region of width D is being used,

$$W(\mathbf{r}_2 - \mathbf{r}_c) = A \operatorname{rect}\left(\frac{x_2 - x_c}{D}\right) \operatorname{rect}\left(\frac{y_2 - y_c}{D}\right) \tag{6.90}$$

(where A is an amplitude factor) so that we have a model point source given by

$$\widetilde{U}_{pt}(\mathbf{r}_1) = A\, e^{-i\frac{k}{2\Delta z}r_1^2} e^{i\frac{k}{2\Delta z}r_c^2} \tag{6.91}$$

$$\times \mathcal{F}^{-1}\left\{\operatorname{rect}\left(\frac{\lambda\Delta z f_x - x_c}{D}\right) \operatorname{rect}\left(\frac{\lambda\Delta z f_y - y_c}{D}\right) e^{-i2\pi\mathbf{r}_c\cdot\mathbf{f}_1}\right\}$$

$$= A\, e^{-i\frac{k}{2\Delta z}r_1^2} e^{i\frac{k}{2\Delta z}r_c^2} e^{-i\frac{k}{\Delta z}\mathbf{r}_c\cdot\mathbf{r}_1} \tag{6.92}$$

$$\times \left(\frac{D}{\lambda\Delta z}\right)^2 \operatorname{sinc}\left[\frac{D(x_1 - x_c)}{\lambda\Delta z}\right] \operatorname{sinc}\left[\frac{D(y_1 - y_c)}{\lambda\Delta z}\right].$$

An example use of a point source is given in Listing 6.8. The point-source model used in the code is shown in Fig. 6.7. The grid spacing is set so that there are ten grid points across the central lobe. This may not seem very point-like, but actually this is only 0.125 mm in diameter. This is much narrower than the window function, which is 8.0 mm across as can be seen in the plot of the propagated irradiance shown in Fig. 6.8. The propagated phase is shown in Fig. 6.9. The effect of the window is clearly visible in both plots, and the model point source is producing exactly what we want in the observation plane region of interest. Later when this model is used for turbulent simulations in Sec. 9.5, the parameter D in the model point source is set to be four times larger than the observing telescope diameter. This ensures that the turbulent fluctuations never cause the window edge to be observed by the telescope.

Unfortunately, Fig. 6.9 does show aliasing outside the region of interest. Perhaps a modification of the point-source model could mitigate some of the aliasing.

Listing 6.9 Example of propagating a sinc-Gaussian model point source in MATLAB using the angular-spectrum method.

```
% example_pt_source_gaussian.m

D = 8e-3;     % diameter of the observation aperture [m]
wvl = 1e-6;   % optical wavelength [m]
k = 2*pi / wvl; % optical wavenumber [rad/m]
Dz = 1;       % propagation distance [m]
arg = D/(wvl*Dz);
delta1 = 1/(10*arg); % source-plane grid spacing [m]
delta2 = D/100; % observation-plane grid spacing [m]
N = 1024;          % number of grid points
% source-plane coordinates
[x1 y1] = meshgrid((-N/2 : N/2-1) * delta1);
[theta1 r1] = cart2pol(x1, y1);
A = wvl * Dz;     % sets field amplitude to 1 in obs plane
pt = A * exp(-i*k/(2*Dz) * r1.^2) * arg^2 ...
    .* sinc(arg*x1) .* sinc(arg*y1) ...
    .* exp(-(arg/4*r1).^2);
[x2 y2 Uout] ...
    = ang_spec_prop(pt, wvl, delta1, delta2, Dz);
```

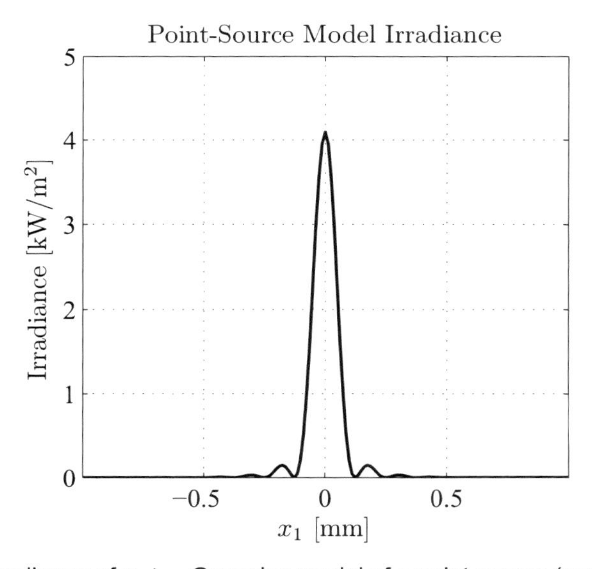

Figure 6.10 Irradiance of a sinc-Gaussian model of a point source (source plane).

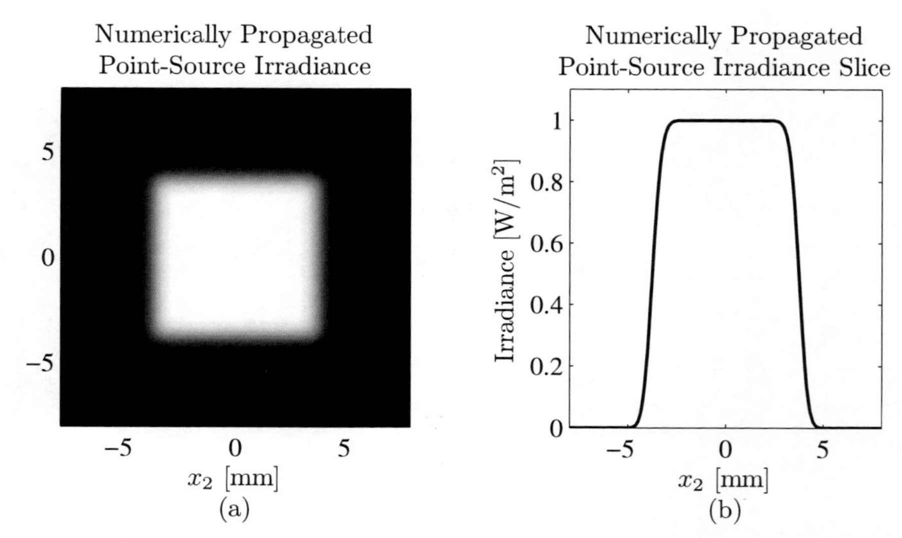

Figure 6.11 Fresnel diffraction irradiance from a sinc-Gaussian model of a point source (observation plane).

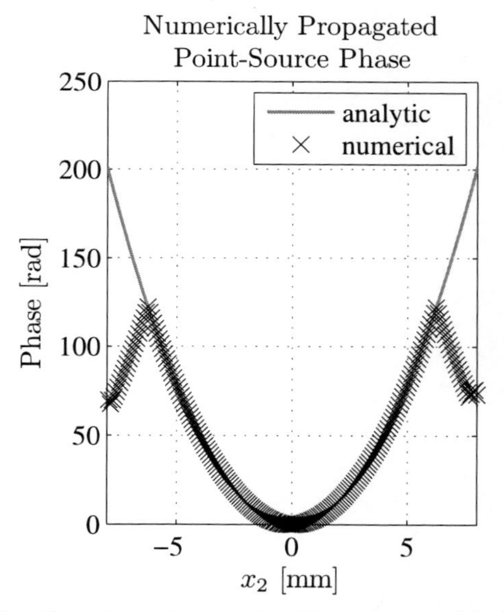

Figure 6.12 Fresnel diffraction phase from a sinc-Gaussian model of a point source (observation plane).

The approaches of Martin and Flatté and Flatté *et al.* do not have such a problem with aliasing because of the Gaussian model they use. Combining the sinc and Gaussian point-source models does, in fact, reduce the phase aliasing slightly. To illustrate, Listing 6.9 implements this. The code is very similar to Listing 6.8, but the model point source is multiplied by a Gaussian function in line 17.

The sinc-Gaussian model point source and resulting observation-plane field are shown in Figs. 6.10–6.12. It is obvious by comparing Figs. 6.7 and 6.10 that the Gaussian factor reduces the side lobes in the model point source and thereby smooths the irradiance profile in the observation-plane field. Further, the computed observation-plane phase shown in Fig. 6.12 matches the analytic phase much better toward the edges of the grid.

6.7 Problems

1. Adjust the example in Listing 6.2 to propagate a Gaussian laser beam using the angular-spectrum method. In the source plane, let the laser beam be at its waist, i.e., $w = w_0 = 1$ mm and $R = \infty$, and let the observation plane be at $z_2 = 4$ m. Use $\lambda = 1$ μm, 512 grid points, a 1-cm grid in the source plane, and a 1.5-cm grid in the observation plane. Show separate plots of the irradiance and phase for the $y_2 = 0$ slice in the observation plane. Include the simulated and analytic results on the same plot for comparison.

2. Adjust the example in Listing 6.2 to propagate a focused beam with a circular aperture using the angular-spectrum method. Let the observation plane be the beam's focal plane. Use $\lambda = 1$ μm, $D = 1$ cm, $f_l = 16$ cm, 1024 grid points, a 2-cm grid in the source plane, and set the grid spacing in the observation plane to be one hundredth of the diffraction-limited spot diameter. Show a plot of the irradiance for the $y_2 = 0$ slice in the focal plane. Include the simulated and analytic results on the same plot for comparison.

3. Adjust the example in Listing 6.2 to simulate Talbot imaging using the angular-spectrum method. Let there be an amplitude grating with amplitude transmittance equal to

$$t_A(x_1, y_1) = \frac{1}{2}\left[1 + \cos\left(2\pi x_1/d\right)\right] \tag{6.93}$$

   in the source plane, and let the observation plane be the first Talbot-image plane. Use $\lambda = 1$ μm, $d = 0.5$ mm, 1024 grid points, a 2 cm grid in both the source plane and observation plane. Show images of the irradiance in the Talbot-image plane (You only need to display the central 10 periods). Display the simulated and analytic results side-by-side for comparison.

4. Compute the model point source if the region of interest is rectangular with widths D_x and D_y in the x_2 and y_2 directions, respectively.

5. Compute the model point source if the region of interest is circular with diameter D.

Chapter 7
Sampling Requirements for Fresnel Diffraction

The primary reason to use simulations is to tackle problems that are analytically intractable. As a result, any computer code that simulates optical-wave propagation needs to handle almost any type of source field. Wave-optics simulations are based on DFTs, and we saw in Ch. 2 that aliasing poses a challenge to DFTs. When the waveform to be transformed is bandlimited, we just need to sample it finely enough to avoid aliasing altogether (satisfying the Nyquist criterion). However, most optical sources are not spatially bandlimited, and the quadratic phase term inside the Fresnel diffraction integral certainly is not bandlimited. These issues have been explored by many authors.[30,31,35,37,42,54,55]

Because an optical field's spatial-frequency spectrum maps directly to its plane-wave spectrum,[5] propagation geometry places a limit on how much spatial frequency content from the source can be seen within the observing aperture. Note that this is physical; it is not caused by sampling. This principle is the foundation of Coy's approach to sampling, and guides most of our discussion on sampling needs in this chapter.

7.1 Imposing a Band Limit

The optical field at each point in the source plane emits a bundle of rays that propagate toward the observation plane. Each ray represents a plane wave propagating in that direction. Let us start by examining the propagation geometry to determine the maximum plane-wave direction relative to the reference normal from the source that is incident upon the region of interest in the observation plane.

Clearly, it is critical to pick the grid spacing and number of grid points to ensure an accurate simulation. The following development uses the propagation geometry to place limits on the necessary spatial-frequency bandwidth, and consequently, the number of sample points and grid spacing. This determines the size and spacing of the source-plane grid and the size and spacing of the observation-plane grid.

At this point, we need to recall the Nyquist criterion to place a constraint on the grid spacing such that

$$\delta \leq \frac{1}{2f_{max}}, \tag{7.1}$$

where f_{max} is the maximum spatial frequency of interest. To build a link between ray angles and spatial bandwidth, we can rewrite Eq. (6.5) in operator notation (just for the FT) as

$$U\left(x_2, y_2\right) = \frac{e^{ik\Delta z}}{i\lambda\Delta z} e^{i\frac{k}{2\Delta z}\left(x_2^2+y_2^2\right)} \mathcal{F}\left[\mathbf{r}_1, \mathbf{f}_1 = \frac{\mathbf{r}_2}{\lambda\Delta z}\right] \left\{U\left(x_1, y_1\right) e^{i\frac{k}{2\Delta z}\left(x_1^2+y_1^2\right)}\right\}. \tag{7.2}$$

The quadratic phase factor inside the FT is interesting; it represents a virtual spherical wave that is focused onto the observation plane. It appears as if the source field's phase is being measured with respect to this spherical surface. After "re-measuring" the phase in this way, the source field is transformed so that each spatial-frequency vector $\mathbf{f}_1$ corresponds to a specific coordinate in the observation plane. Below, we exploit this link between geometry and spatial frequency to levy constraints on the sampling grids.

In the angular-spectrum formulation of diffraction, the concept is that an optical field $U\left(x, y\right)$ may be decomposed into a sum of plane waves with varying amplitudes and directions. A plane wave $U_p\left(x, y, z, t\right)$ with arbitrary direction is given in phasor notation by

$$U_p\left(x, y, z, t\right) = e^{i(\mathbf{k}\cdot\mathbf{r}-2\pi\nu t)}, \tag{7.3}$$

where $\mathbf{r} = x\hat{\mathbf{i}} + y\hat{\mathbf{j}} + z\hat{\mathbf{k}}$ is a three-dimensional position vector, $\mathbf{k} = (2\pi/\lambda)\left(\alpha\hat{\mathbf{i}} + \beta\hat{\mathbf{j}} + \gamma\hat{\mathbf{k}}\right)$ is the optical wavevector, and ν is the temporal frequency of the optical wave. These direction cosines are depicted in Fig. 7.1. Using phasor notation, a plane wave is given by

$$U_p\left(x, y, z\right) = e^{i\mathbf{k}\cdot\mathbf{r}} = e^{i\frac{2\pi}{\lambda}(\alpha x+\beta y)} e^{i\frac{2\pi}{\lambda}\gamma z}. \tag{7.4}$$

In the $z = 0$ plane, a complex-exponential source in the form $\exp\left[i2\pi\left(f_x x + f_y y\right)\right]$ may be regarded as a plane wave propagating with direction cosines

$$\alpha = \lambda f_x, \qquad \beta = \lambda f_y, \qquad \gamma = \sqrt{1 - (\lambda f_x)^2 - (\lambda f_y)^2}. \tag{7.5}$$

Therefore, the spatial-frequency spectrum of an optical source is also its plane-wave spectrum with the spatial frequencies mapped to direction cosines (α, β), where the mapping is given in Eqs. (7.5). Figure 7.1 illustrates the geometry of these direction cosines. From this, the angular spectrum's cutoff angle is defined as $\alpha_{max} = \lambda f_{max}$, where α_{max} is the maximum angle in the angular spectrum that can affect the observed field. Now, Eq. (7.1) may be rewritten to relate an optical field's maximum angular content to the grid spacing so that

$$\delta_1 \leq \frac{\lambda}{2\alpha_{max}}. \tag{7.6}$$

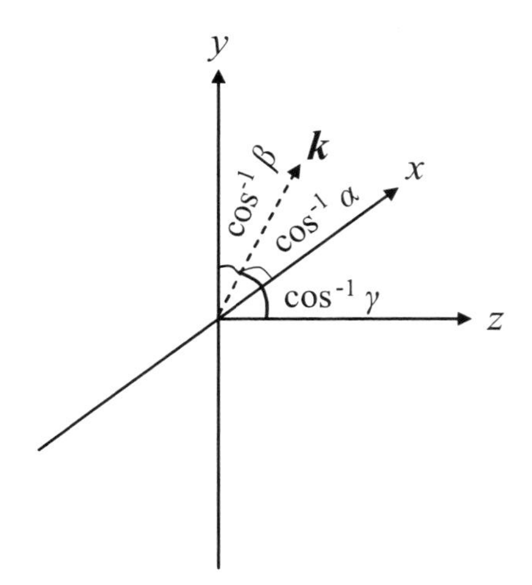

Figure 7.1 Depiction of direction cosines α, β, and γ.

Conversely, if the grid spacing is given, then the maximum angular content represented by the sampled version of the optical field is

$$\alpha_{max} = \frac{\lambda}{2\delta_1}. \tag{7.7}$$

This allows us to tie grid parameters to the propagation geometry.

7.2 Propagation Geometry

Now, the task is to use the sizes of the source and receiver to determine α_{max}. This section follows the developments of Coy, Praus, and Mansell.[35,42,54] The discussion is restricted to one spatial dimension, but it may easily be generalized to two dimensions. Additionally, the propagating wavefront is assumed to be spherical for generality.

As shown in Fig. 7.2, the source field has a maximum spatial extent D_1. In the observation plane, the region of interest has a maximum spatial extent D_2. Perhaps the optical field is propagating to a sensor, and D_2 is the diameter of the sensor. Additionally, let the grid spacing in the source plane be δ_1 and the grid spacing in the observation plane be δ_2.

While the source field can be considered a sum of plane waves as discussed above, it can alternately be considered a sum of point sources. This is precisely Huygens' principle. We take this view so that we ensure the grids are sampled finely enough that each point in the source field fully illuminates the observation-plane region of interest. The maximum ray angle α_{max} corresponds to the divergence angle of source-plane field points.

Consider a point at the lower edge of the source, at point $(x_1 = -D_1/2, z = z_1)$. The angle α_{max} can be written as the sum of two angles α_k and α_{edges}, as shown in Fig. 7.2. The angle between the bottom edge of the source and the top edge of

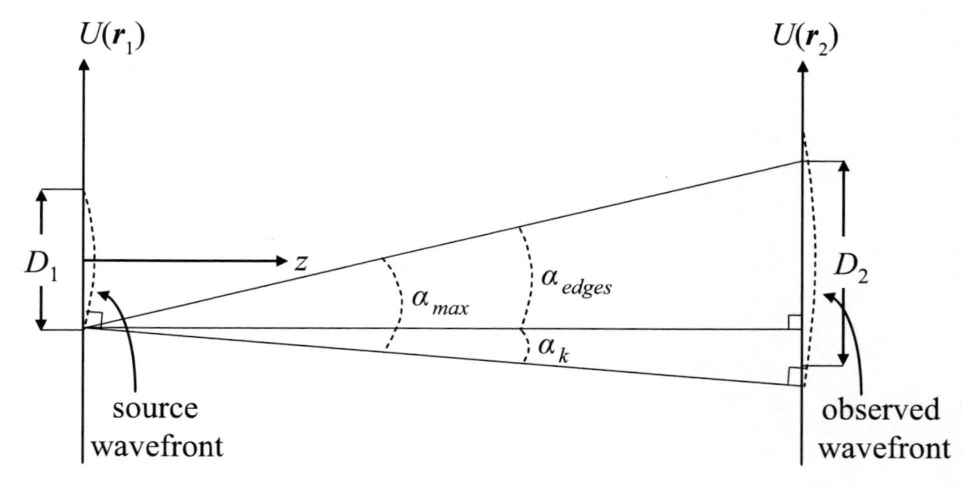

Figure 7.2 Definition of angles α_{max}, α_{edges}, and α_k.

the observing aperture, at point $(x_2 = D_2/2, z = z_2)$, is (in the paraxial approximation)

$$\alpha_{edges} = \frac{D_1 + D_2}{2\Delta z}. \tag{7.8}$$

At the lower edge of the source, the optical wavevector $\mathbf{k}$ of the virtual spherical wave apparent in Eq. (7.2) makes an angle with the z axis. Because there is a fixed number of grid points, spaced by a distance δ_1 in the source plane and δ_2 in the observation plane, the ratio of the grid sizes (observation/source) is δ_2/δ_1. Thus, $\mathbf{k}$ intersects the observation plane at $x_2 = -D_1\delta_2/(2\delta_1)$. Consequently, the (paraxial) angle α_k is given by

$$\alpha_k = \frac{D_1\delta_2}{2\delta_1\Delta z} - \frac{D_1}{2\Delta z} = \frac{D_1}{2\Delta z}\left(\frac{\delta_2}{\delta_1} - 1\right). \tag{7.9}$$

Then, α_{max} is given by

$$\alpha_{max} = \alpha_{edges} + \alpha_k \tag{7.10}$$

$$= \frac{D_1 + D_2}{2\Delta z} + \frac{D_1}{2\Delta z}\left(\frac{\delta_2}{\delta_1} - 1\right) \tag{7.11}$$

$$= \frac{D_1\,\delta_2/\delta_1 + D_2}{2\Delta z}. \tag{7.12}$$

When this is combined with the sampling requirement in Eq. (7.7), the result is

$$\frac{D_1\,\delta_2/\delta_1 + D_2}{2\Delta z} \le \frac{\lambda}{2\delta_1} \tag{7.13}$$

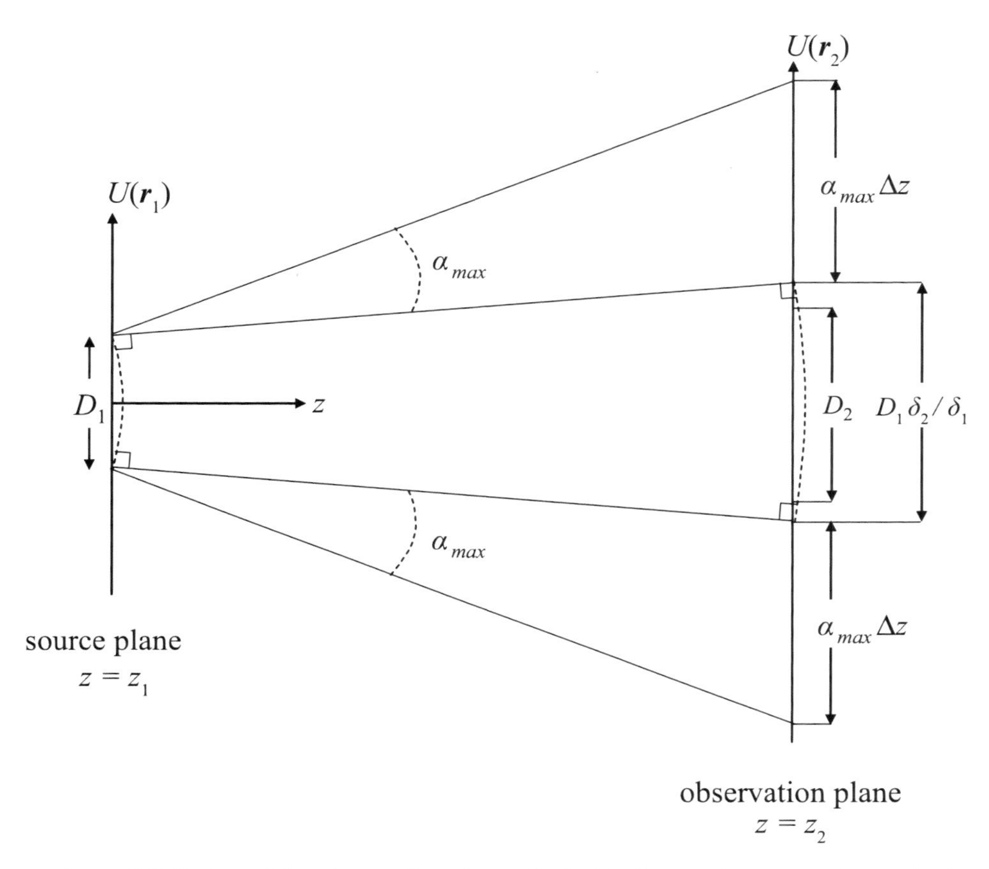

Figure 7.3 Portion of the observation plane affected by the maximum angular content.

$$\delta_2 \leq -\frac{D_2}{D_1}\delta_1 + \frac{\lambda\Delta z}{D_1}. \tag{7.14}$$

Satisfying Eq. (7.14) means that the selected grid spacings adequately sample the spatial bandwidth that affects the observation-plane region of interest.

Now, it is useful to determine the necessary spatial extent of the observation-plane grid. Figure 7.3 shows that the diameter D_{illum} of illuminated area (by a source with maximum angular content α_{max}) in the observation plane is

$$D_{illum} = D_1\delta_2/\delta_1 + 2\alpha_{max}\Delta z \tag{7.15}$$

$$= D_1\delta_2/\delta_1 + \frac{\lambda\Delta z}{\delta_1}. \tag{7.16}$$

Aliasing in the observation plane is allowable as long as it does not invade the area of the observing aperture. If the grid has a smaller spatial extent than the illuminated area, we can imagine the edges of the illuminated area wrapping around to the other side of the grid. Recall that this is apparent in Figs. 2.6(d) and 2.7(d). For the wrapping to get just to the edge of the observing aperture, the grid extent must be at least as large as the mean of the illuminated area and the observing aperture diameter so that it wraps only half-way around, yielding

$$D_{grid} \geq \frac{D_{illum} + D_2}{2} \tag{7.17}$$

$$= \frac{D_1 \delta_2/\delta_1 + \lambda \Delta z/\delta_1 + D_2}{2}. \tag{7.18}$$

Finally, the number of grid points required in the observation plane is

$$N = \frac{D_{grid}}{\delta_2} \tag{7.19}$$

$$\geq \frac{D_1}{2\delta_1} + \frac{D_2}{2\delta_2} + \frac{\lambda \Delta z}{2\delta_1 \delta_2}. \tag{7.20}$$

Satisfying Eq. (7.20) means that the spatial extent of the observation plane is large enough to ensure that the light that wraps around does not creep into the observation-plane region of interest.

7.3 Validity of Propagation Methods

Unfortunately, satisfying the geometric constraints to avoid aliasing in the observation-plane region of interest does not guarantee satisfactory results. One must also consider which method of propagation can be used. The Fresnel-integral method and the angular-spectrum method have different constraints. One must avoid aliasing the quadratic phase factor inside the FTs that are used, and the two propagation methods have different two quadratic phase factors. With these different constraints, it turns out that the Fresnel-integral approach from Sec. 6.3 is valid for long propagations, while the angular-spectrum approach from Sec. 6.4 is valid for short propagations.[30,31,37]

7.3.1 Fresnel-integral propagation

This subsection begins by applying the geometric constraints with consideration of the particular grid spacing allowed by Fresnel-integral propagation. Then, it goes on to examine how to avoid aliasing of the quadratic phase factor in the source plane. These analyses result in a set of inequalities that must be satisfied when choosing the grid parameters.

7.3.1.1 One step, fixed observation-plane grid spacing

As discussed in the previous chapter, the observation-plane grid spacing δ_2 is fixed when one executes a single step of Fresnel-integral propagation. This fixed value is

$$\delta_2 = \frac{\lambda \Delta z}{N \delta_1}. \tag{7.21}$$

Relating this to the propagation geometry, we substitute this into Eq. (7.14), which yields

$$D_1 \frac{\lambda \Delta z}{N \delta_1} + D_2 \delta_1 \leq \lambda \Delta z \tag{7.22}$$

$$D_1 \frac{\lambda \Delta z}{\delta_1} + D_2 \delta_1 N \leq N \lambda \Delta z \tag{7.23}$$

$$D_1 \frac{\lambda \Delta z}{\delta_1} \leq N \left(\lambda \Delta z - D_2 \delta_1 \right) \tag{7.24}$$

$$N \geq \frac{D_1 \lambda \Delta z}{\delta_1 \left(\lambda \Delta z - D_2 \delta_1 \right)}. \tag{7.25}$$

Substituting for δ_2 in Eq. (7.20) yields

$$N \geq \frac{D_1}{2\delta_1} + \frac{D_2 \delta_1}{2\lambda \Delta z} N + \frac{\lambda \Delta z}{2\delta_1} \frac{N \delta_1}{\lambda \Delta z} \tag{7.26}$$

$$N \geq \frac{D_1}{2\delta_1} + \frac{D_2 \delta_1}{2\lambda \Delta z} N + \frac{N}{2} \tag{7.27}$$

$$\frac{N}{2} - \frac{D_2 \delta_1}{2\lambda \Delta z} N \geq \frac{D_1}{2\delta_1} \tag{7.28}$$

$$N \left(1 - \frac{D_2 \delta_1}{\lambda \Delta z} \right) \geq \frac{D_1}{\delta_1} \tag{7.29}$$

$$N \geq \frac{D_1}{\delta_1 \left(1 - \frac{D_2 \delta_1}{\lambda \Delta z} \right)} \tag{7.30}$$

$$N \geq \frac{D_1 \lambda \Delta z}{\delta_1 \left(\lambda \Delta z - D_2 \delta_1 \right)}. \tag{7.31}$$

This is identical to Eq. (7.25)! Also notice two properties of this inequality: we must have $\lambda \Delta z > D_2 \delta_1$ because N can only be positive, and as $\lambda \Delta z \to D_2 \delta_1$ the minimum necessary N approaches ∞.

7.3.1.2 Avoiding aliasing

The free-space amplitude spread function has a very large bandwidth. In fact, the cutoff frequency is λ^{-1}, which is impractically high to represent on a grid of finite size.[5] If we tried to use a source-plane grid spacing of $\delta_1 = \lambda/2 \approx 500$ nm, the largest grid extent that could be used is $L = N\delta_1 \approx 500$ nm $\times 1024 = 0.512$ mm (grid sizes up to 2048 or 4096 might be possible, depending on the computer being used). Of course, very few practical problems can be simulated on such a small grid.

In practice, the best one can do is to ensure that all of the frequencies present on the grid are represented correctly. We cannot plan for all possible kinds of source-plane fields, so we derive a sampling guideline by modeling the source as an apodized beam with maximum spatial extent D_1 and a parabolic wavefront with radius R. This source field $U(\mathbf{r}_1)$ can be written as

$$U(\mathbf{r}_1) = A(\mathbf{r}_1)\, e^{i\frac{k}{2R}r_1^2}, \tag{7.32}$$

where $A(\mathbf{r}_1)$ describes the amplitude transmittance of the source aperture. The maximum spatial extent of the nonzero portions of $A(\mathbf{r}_1)$ is D_1. A diverging beam is indicated by $R < 0$, while a converging beam is indicated by $R > 0$. With this type of source, the Fresnel diffraction integral becomes

$$U(\mathbf{r}_2) = \mathcal{Q}\left[\frac{1}{\Delta z}, \mathbf{r}_2\right] \mathcal{V}\left[\frac{1}{\lambda\Delta z}, \mathbf{r}_1\right] \mathcal{F}[\mathbf{r}_1, \mathbf{f}_1]\, \mathcal{Q}\left[\frac{1}{\Delta z}, \mathbf{r}_1\right] \{U(\mathbf{r}_1)\} \tag{7.33}$$

$$= \mathcal{Q}\left[\frac{1}{\Delta z}, \mathbf{r}_2\right] \mathcal{V}\left[\frac{1}{\lambda\Delta z}, \mathbf{r}_1\right] \mathcal{F}[\mathbf{r}_1, \mathbf{f}_1]\, \mathcal{Q}\left[\frac{1}{\Delta z}, \mathbf{r}_1\right] \left\{A(\mathbf{r}_1)\, e^{i\frac{k}{2R}r_1^2}\right\} \tag{7.34}$$

$$= \mathcal{Q}\left[\frac{1}{\Delta z}, \mathbf{r}_2\right] \mathcal{V}\left[\frac{1}{\lambda\Delta z}, \mathbf{r}_1\right] \mathcal{F}[\mathbf{r}_1, \mathbf{f}_1]\, \mathcal{Q}\left[\frac{1}{\Delta z}, \mathbf{r}_1\right] \mathcal{Q}\left[\frac{1}{R}, \mathbf{r}_1\right] \{A(\mathbf{r}_1)\} \tag{7.35}$$

$$= \mathcal{Q}\left[\frac{1}{\Delta z}, \mathbf{r}_2\right] \mathcal{V}\left[\frac{1}{\lambda\Delta z}, \mathbf{r}_1\right] \mathcal{F}[\mathbf{r}_1, \mathbf{f}_1]\, \mathcal{Q}\left[\frac{1}{\Delta z} + \frac{1}{R}, \mathbf{r}_1\right] \{A(\mathbf{r}_1)\}. \tag{7.36}$$

The key to achieving an accurate result is to sample the quadratic phase factor inside the FT at a high enough rate to satisfy the Nyquist criterion. If it is not sampled finely enough, the intended high-frequency content would show up in the lower frequencies. Again, this effect is visible in Figs. 2.6(d) and 2.7(d). Lower frequencies map to lower ray angles that may erroneously impinge on the observation-plane region of interest.

To avoid or at least minimize aliasing, we need to determine the bandwidth of the product $\mathcal{Q}A$ from Eq. (7.36). Lambert and Fraser demonstrated that for very small apertures, the bandwidth is set by A, while for larger apertures, it is set by the phase of $\mathcal{Q}$ at the edge of the aperture.[47] Typically, the latter is the case, so we focus on the phase of $\mathcal{Q}$. Local spatial frequency $\mathbf{f}_{loc}$ is basically the local rate of change of a waveform given by[5]

$$\mathbf{f}_{loc} = \frac{1}{2\pi}\nabla\phi, \tag{7.37}$$

where ϕ is the optical phase measured in radians, and the Cartesian components of $\mathbf{f}_{loc}$ are measured in m^{-1}. Conceptually, a waveform with rapid variations (regions of large gradients) has high-frequency content. We want to find the maximum local spatial frequency of the quadratic phase factor inside the integral and sample at least twice this rate. Since the quadratic phase has the same variations in the both

Cartesian directions, we just analyze the x_1 direction, which yields

$$f_{locx} = \frac{1}{2\pi}\frac{\partial}{\partial x_1}\frac{k}{2}\left(\frac{1}{\Delta z}+\frac{1}{R}\right)r_1^2 \tag{7.38}$$

$$= \left(\frac{1}{\Delta z}+\frac{1}{R}\right)\frac{x_1}{\lambda}. \tag{7.39}$$

This takes on its maximum value at the edge of the grid where $x_1 = N\delta_1/2$. However, if the source is apodized, and the field is nonzero only within a centered aperture of maximum extent D_1, then that includes the phase. Thus, the product of the source field and the quadratic phase factor has its maximum local spatial frequency value at $x_1 = \pm D_1/2$. Then, applying the Nyquist criterion yields

$$\left(\frac{1}{\Delta z}+\frac{1}{R}\right)\frac{D_1}{2\lambda} \leq \frac{1}{2\delta_1}. \tag{7.40}$$

After some algebra, we obtain

$$\Delta z \geq \frac{D_1\delta_1 R}{\lambda R - D_1\delta_1} \qquad \text{for finite } R \tag{7.41}$$

$$\Delta z \geq \frac{D_1\delta_1}{\lambda} \qquad \text{for infinite } R. \tag{7.42}$$

Note that this is just a guideline. When Δz is close to its minimum required value, the simulation results may not match analytic results perfectly.

The following example illustrates the process of using a sound analysis of sampling to obtain accurate simulation results. Listing 7.1 gives an example of subsequent usage of `one_step_prop` for a square aperture with due consideration of sampling constraints. It goes on to plot the results along with the analytic result. In line 10, the minimum number of grid points is computed using Eq. (7.25). In this example, 66 grid points are required. Then, in line 11, the number of grid points to actually use is determined by using the next power of two, which is 128. This is done to take advantage of the FFT algorithm. After line 11 executes, the sampling-related parameters for this simulation are

$$\begin{aligned} D_1 &= 2\,\text{mm} \\ D_2 &= 3\,\text{mm} \\ \lambda &= 1\,\mu\text{m} \\ \Delta z &= 0.5\,\text{m} \\ \delta_1 &= 40\,\mu\text{m} \\ \delta_2 &= 97.7\,\mu\text{m} \\ N &= 128. \end{aligned} \tag{7.43}$$

Applying Eq. (7.42), we find that the minimum distance to use one step of Fresnel-integral propagation is 8 cm. Clearly, we can expect results that match theory

Listing 7.1 Example of evaluating the Fresnel diffraction integral in MATLAB using a single step.

```
1  % example_square_one_step_prop_samp.m
2
3  D1 = 2e-3;   % diam of the source aperture [m]
4  D2 = 3e-3;   % diam of the obs-plane region of interest [m]
5  delta1 = D1 / 50;     % want at least 50 grid pts across ap
6  wvl = 1e-6;   % optical wavelength [m]
7  k = 2*pi / wvl;
8  Dz = 0.5;            % propagation distance [m]
9  % minimum number of grid points
10 Nmin = D1 * wvl*Dz / (delta1 * (wvl*Dz - D2*delta1));
11 N = 2^ceil(log2(Nmin));      % number of grid pts per side
12 % source plane
13 [x1 y1] = meshgrid((-N/2 : N/2-1) * delta1);
14 ap = rect(x1/D1) .* rect(y1/D1);
15 % simulate the propagation
16 [x2 y2 Uout] = one_step_prop(ap, wvl, delta1, Dz);
17
18 % analytic result for y2=0 slice
19 Uout_an ...
20     = fresnel_prop_square_ap(x2(N/2+1,:), 0, D1, wvl, Dz);
```

closely because there are more than enough grid points (by nearly a factor of two), and the propagation is much farther than the limit required by this simulation method. Figure 7.4 shows the resulting amplitude and phase. The simulation does, in fact, match the analytic results closely.

7.3.2 Angular-spectrum propagation

For the angular-spectrum method, the observation-plane grid spacing is not fixed like in the previous section. The grid spacings δ_1 and δ_2 can be chosen independently so, there are no simplifications to Eqs. (7.14) and (7.20) like with the Fresnel-integral method. Instead, there are two additional inequalities that must be satisfied to keep high-frequency content from corrupting the observation-plane region of interest. This is because the angular-spectrum method from Eq. (6.67) has its own requirements to avoid aliasing of a quadratic phase factor. As in the previous section, we restrict the source-plane field $U(\mathbf{r}_1)$ to the form in Eq. (7.32). With this form, the angular-spectrum method can be written as

$$U(\mathbf{r}_2) = \mathcal{Q}\left[\frac{m-1}{m\Delta z}, \mathbf{r}_2\right] \mathcal{F}^{-1}\left[\mathbf{f}_1, \frac{\mathbf{r}_2}{m}\right] \mathcal{Q}_2\left[-\frac{\Delta z}{m}, \mathbf{f}_1\right] \times \mathcal{F}[\mathbf{r}_1, \mathbf{f}_1]\, \mathcal{Q}\left[\frac{1-m}{\Delta z}, \mathbf{r}_1\right] \frac{1}{m}\{U(\mathbf{r}_1)\} \tag{7.44}$$

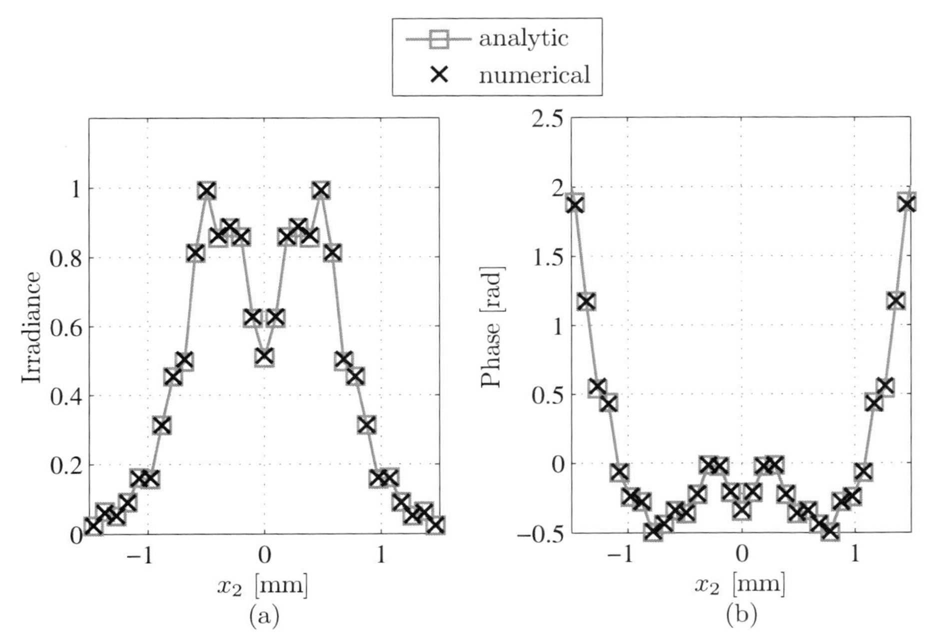

Figure 7.4 Fresnel diffraction from a square aperture, simulation and analytic: (a) observation-plane irradiance and (b) observation-plane phase.

$$
\begin{aligned}
&= \mathcal{Q}\left[\frac{m-1}{m\Delta z}, \mathbf{r}_2\right] \mathcal{F}^{-1}\left[\mathbf{f}_1, \frac{\mathbf{r}_2}{m}\right] \mathcal{Q}_2\left[-\frac{\Delta z}{m}, \mathbf{f}_1\right] \\
&\quad \times \mathcal{F}\left[\mathbf{r}_1, \mathbf{f}_1\right] \mathcal{Q}\left[\frac{1-m}{\Delta z}, \mathbf{r}_1\right] \frac{1}{m}\left\{A\left(\mathbf{r}_1\right) e^{i\frac{k}{2R}r_1^2}\right\} \\
&= \mathcal{Q}\left[\frac{m-1}{m\Delta z}, \mathbf{r}_2\right] \mathcal{F}^{-1}\left[\mathbf{f}_1, \frac{\mathbf{r}_2}{m}\right] \mathcal{Q}_2\left[-\frac{\Delta z}{m}, \mathbf{f}_1\right] \\
&\quad \times \mathcal{F}\left[\mathbf{r}_1, \mathbf{f}_1\right] \mathcal{Q}\left[\frac{1-m}{\Delta z}, \mathbf{r}_1\right] \frac{1}{m} \mathcal{Q}\left[\frac{1}{R}, \mathbf{r}_1\right] \left\{A\left(\mathbf{r}_1\right)\right\} \\
&= \mathcal{Q}\left[\frac{m-1}{m\Delta z}, \mathbf{r}_2\right] \mathcal{F}^{-1}\left[\mathbf{f}_1, \frac{\mathbf{r}_2}{m}\right] \mathcal{Q}_2\left[-\frac{\Delta z}{m}, \mathbf{f}_1\right] \\
&\quad \times \mathcal{F}\left[\mathbf{r}_1, \mathbf{f}_1\right] \frac{1}{m} \mathcal{Q}\left[\frac{1-m}{\Delta z} + \frac{1}{R}, \mathbf{r}_1\right] \left\{A\left(\mathbf{r}_1\right)\right\}.
\end{aligned}
\tag{7.45}
$$

There are two quadratic phase factors inside the FT (and IFT) operations to consider:

$$
\mathcal{Q}\left[\frac{1-m}{\Delta z} + \frac{1}{R}, \mathbf{r}_1\right] = \exp\left[-i\frac{k}{2}\left(\frac{1-m}{\Delta z} + \frac{1}{R}\right) \left|\mathbf{r}_1\right|^2\right] \tag{7.46}
$$

$$
\mathcal{Q}_2\left[-\frac{\Delta z}{m}, \mathbf{f}_1\right] = \exp\left(i\pi^2 \frac{2\Delta z}{mk} \left|\mathbf{f}_1\right|^2\right). \tag{7.47}
$$

Like in the previous section, we need to compute the maximum local spatial frequency in each factor and apply the Nyquist sampling criterion. This ensures that all of the present spatial frequencies are not aliased, thus preserving the observation-plane field within the region of interest.

In the first phase factor, the phase ϕ is

$$\phi = \frac{k}{2}\left(\frac{1-m}{\Delta z} + \frac{1}{R}\right)|\mathbf{r}_1|^2 \tag{7.48}$$

$$= \frac{k}{2}\left(\frac{1-\delta_2/\delta_1}{\Delta z} + \frac{1}{R}\right)|\mathbf{r}_1|^2 . \tag{7.49}$$

The local spatial frequency f_{lx} is

$$f_{lx} = \frac{1}{2\pi}\frac{\partial}{\partial x_1}\phi \tag{7.50}$$

$$= \frac{1}{\lambda}\left(\frac{1-\delta_2/\delta_1}{\Delta z} + \frac{1}{R}\right)x_1 . \tag{7.51}$$

Once again, the maximum spatial frequency occurs at $x_1 = \pm D_1/2$ because this factor is multiplied by the source-plane pupil function. Applying the Nyquist sampling gives

$$\frac{1}{\lambda}\left|\frac{1-\delta_2/\delta_1}{\Delta z} + \frac{1}{R}\right|\frac{D_1}{2} \leq \frac{1}{2\delta_1} . \tag{7.52}$$

After some algebra, we obtain

$$\left(1+\frac{\Delta z}{R}\right)\delta_1 - \frac{\lambda\Delta z}{D_1} \leq \delta_2 \leq \left(1+\frac{\Delta z}{R}\right)\delta_1 + \frac{\lambda\Delta z}{D_1} . \tag{7.53}$$

The phase of the second quadratic phase factor (the amplitude transfer function) is

$$\phi = \pi^2\frac{2\Delta z}{mk}|\mathbf{f}_1|^2 \tag{7.54}$$

$$= \pi^2\frac{2\delta_1\Delta z}{\delta_2 k}|\mathbf{f}_1|^2 . \tag{7.55}$$

The local spatial frequency f'_{lx} (prime notation to avoid confusion with the variable in the quadratic phase factor) is

$$f'_{lx} = \frac{1}{2\pi}\frac{\partial}{\partial f_{1x}}\phi \tag{7.56}$$

$$= \frac{\delta_1\lambda\Delta z}{\delta_2} f_{1x} . \tag{7.57}$$

This is a maximum at the edge of the spatial-frequency grid where $f_{1x} = \pm 1/(2\delta_1)$. Applying Nyquist sampling criterion gives

$$\frac{\lambda\Delta z}{2\delta_2} \leq \frac{N\delta_1}{2} \tag{7.58}$$

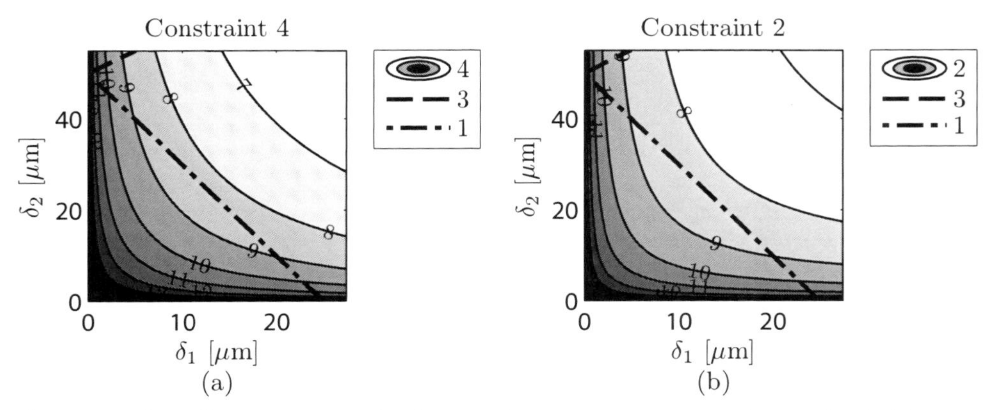

Figure 7.5 Sampling constraints for the angular-spectrum propagation method: (a) constraints 4, 3, and 1; (b) constraints 2, 3, and 1.

$$N \geq \frac{\lambda \Delta z}{\delta_1 \delta_2}. \tag{7.59}$$

Because there are four inequalities, the procedure here is more complicated than for Fresnel-integral propagation. Again, the simplest way to illustrate this procedure is by example. Let us restate the sampling constraints grouped together:

1. $\delta_2 \leq -\frac{D_2}{D_1}\delta_1 + \frac{\lambda \Delta z}{D_1}$,

2. $N \geq \frac{D_1}{2\delta_1} + \frac{D_2}{2\delta_2} + \frac{\lambda \Delta z}{2\delta_1 \delta_2}$,

3. $\left(1 + \frac{\Delta z}{R}\right)\delta_1 - \frac{\lambda \Delta z}{D_1} \leq \delta_2 \leq \left(1 + \frac{\Delta z}{R}\right)\delta_1 + \frac{\lambda \Delta z}{D_1}$,

4. $N \geq \frac{\lambda \Delta z}{\delta_1 \delta_2}$.

Consider an example of evaluating Eq. (7.44) for the following parameters: $D_1 = 2$ mm, $D_2 = 4$ mm, $\Delta z = 0.1$ m, and $\lambda = 1\,\mu$m. Solving four inequalities simultaneously is challenging. The simplest approach is to graphically display the bounds for these inequalities in the (δ_1, δ_2) domain. These are shown in Fig. 7.5. Plot (a) shows a contour plot of the lower bound on $\log_2 N$ from constraint 4 (solid black lines). Also on the plot are the upper bounds on δ_2 given by constraints 1 (dash-dot line) and 3 (dashed line barely visible in the upper-left corner). Constraint 1 is clearly more restrictive than constraint 3 where δ_2 is concerned. When choosing values for δ_1 and δ_2, this limits us to the lower-left corner of the plot below the dotted line. The required number of grid points in this region of the contour plot is at least $2^{8.5}$. However, we realistically must pick an integer power of two to take advantage of the FFT algorithm, so it looks like we must choose $N = 2^9 = 512$ grid points. Somewhat arbitrarily choosing $\delta_1 = 9.48\,\mu$m and $\delta_2 = 28.12\,\mu$m, the minimum required number of grid points is $2^{8.55}$. Consequently, we must choose $N = 2^9 = 512$ grid points unless constraint 2 is more restrictive. Plot (b) indicates

Listing 7.2 Example of evaluating the Fresnel diffraction integral in MATLAB using the angular-spectrum method.

```
1  % example_square_prop_ang_spec.m
2
3  D1 = 2e-3;     % diameter of the source aperture [m]
4  D2 = 4e-3;     % diameter of the observation aperture [m]
5  wvl = 1e-6;  % optical wavelength [m]
6  k = 2*pi / wvl;
7  Dz = 0.1;         % propagation distance [m]
8  delta1 = 9.4848e-6;
9  delta2 = 28.1212e-6;
10 Nmin = D1/(2*delta1) + D2/(2*delta2) ...
11     + (wvl*Dz)/(2*delta1*delta2);
12 % bump N up to the next power of 2 for efficient FFT
13 N = 2^ceil(log2(Nmin));
14
15 [x1 y1] = meshgrid((-N/2 : N/2-1) * delta1);
16 ap =rect(x1/D1) .* rect(y1/D1);
17 [x2 y2 Uout] = ang_spec_prop(ap, wvl, delta1, delta2, Dz);
18
19 % analytic result for y2=0 slice
20 Uout_an ...
21     = fresnel_prop_square_ap(x2(N/2+1,:), 0, D1, wvl, Dz);
```

that the required number of grid points according to constraint 2 is only $2^{8.51}$. As a result, picking $N = 512$ is sufficient, given that $\delta_1 = 9.48\,\mu$m and $\delta_2 = 28.12\,\mu$m.

Listing 7.2 gives the MATLAB code for the simulation in this example. The code numerically evaluates the angular-spectrum method [Eq. (7.44)] to simulate propagation from a square aperture. The simulation uses the parameters from this discussion of sampling. Given all of this consideration to sampling, one expects that the amplitude and phase of the simulated result should match the analytic results closely. These results are shown in Fig. 7.6 with a $y_2 = 0$ slice of the irradiance shown in plot (a) and a $y_2 = 0$ slice of the wrapped phase shown in plot (b). Indeed, the simulation result does match the analytic result closely.

7.3.3 General guidelines

We can now formulate this problem more generally. First, it can be shown that constraint 4 is more restrictive than the combination of constraints 1 and 2. Therefore, only Fig. 7.5(a) needs to be analyzed, and plot (b) may be ignored. Further, constraints 2 and 3 are simple linear inequalities. Constraint 1 has a slope of $-D_2/D_1$ and a δ_2-intercept of $\lambda\Delta z/D_1$, as shown in Fig. 7.7. Constraint 3 is more interesting, however. The upper bound has a slope of $1 + \Delta z/R$ and a δ_2-

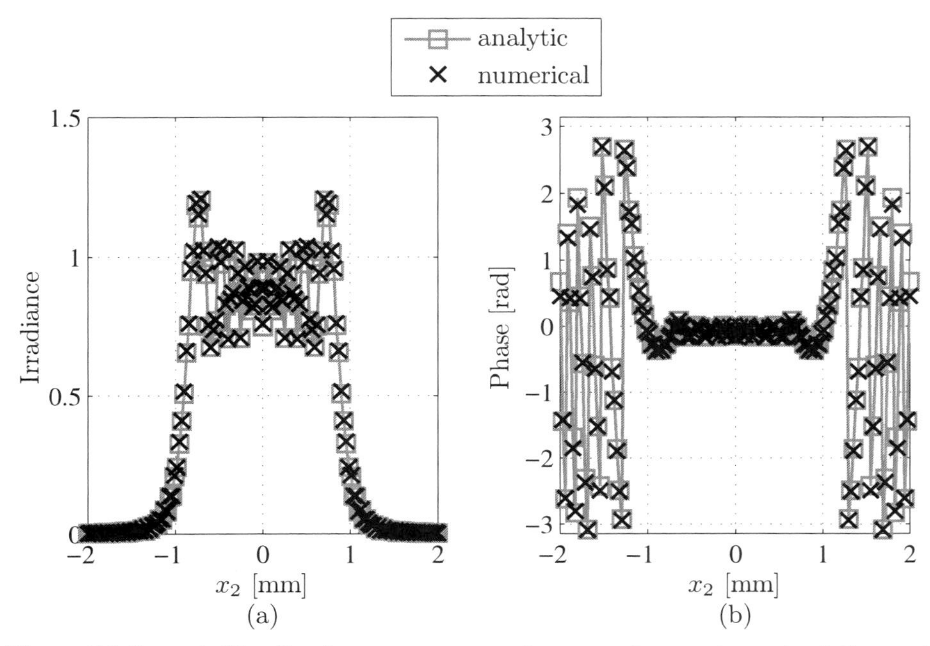

Figure 7.6 Fresnel diffraction from a square aperture, angular-spectrum simulation and analytic: (a) observation-plane irradiance and (b) observation-plane phase.

intercept of $\lambda\Delta z/D_1$. Comparing Fig. 7.7 (a) and (b) with plot (c) shows that if $-D_2/D_1 < 1 + \Delta z/R$, the upper bound on constraint 3 is not a consideration because it has the same δ_2-intercept and a greater slope than constraint 2. The lower bound of constraint 3 has a slope of $1 + \Delta z/R$ and a δ_2-intercept of $-\lambda\Delta z/D_1$. The δ_2-intercept is unphysical, so we disregard it and instead focus on the δ_1 intercept, which is $\lambda\Delta z/\left[D_1\left(1 + \Delta z/R\right)\right]$. Therefore, comparing plots (a) and (b) reveals that when $1 + \Delta z/R < D_2/D_1$, the lower bound of constraint 3 is not a factor.

To summarize the above discussion of constraint 3, when

$$\left|1 + \frac{\Delta z}{R}\right| < \frac{D_2}{D_1}, \tag{7.60}$$

constraint 3 is not a factor. Interestingly, the physical interpretation is that the geometric beam is contained within a region of diameter D_2. This includes diverging source fields and converging source fields that are focused in front of and behind the observation plane.

This analysis of sampling constraints should serve as a guideline for wave-optics simulations, but not as unbreakable rules. The most important lesson from this chapter is that quadratic phase factors, which are ubiquitous in Fourier optics, pose great challenges to numerical evaluation, so simulations must be approached carefully and validated fully. When attempting to simulate a Fourier-optics propagation problem that does not have a known analytic solution, one must consider

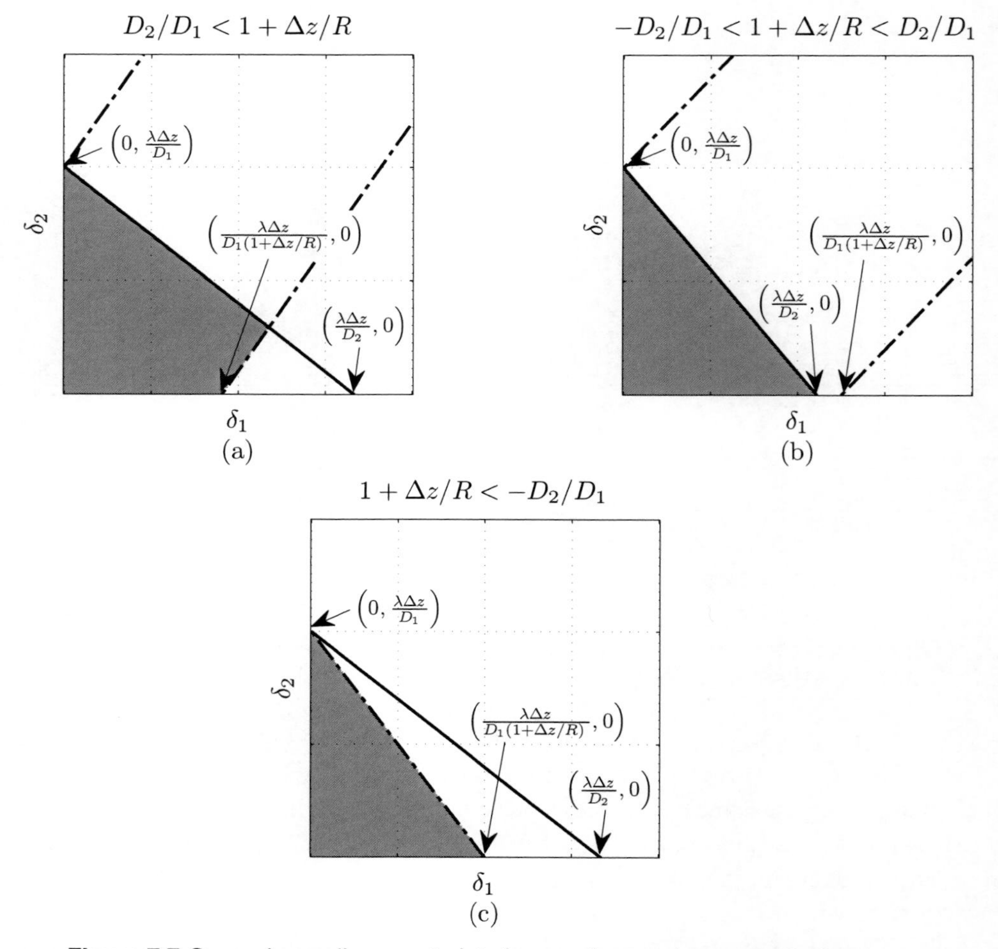

Figure 7.7 General sampling constraints for angular-spectrum propagation.

sampling first as a general guideline for choosing the propagation grids. Then, the accuracy of the simulation setup must be validated through the simulation of a similar problem with a known solution. That is why this book makes such heavy use of the square-aperture propagation problem.

7.4 Problems

1. Consider the signal

$$g(x) = \exp\left(i\pi a^2 x^2\right) \tag{7.61}$$

with $a = 4$ sampled on a grid with $N = 128$ points and $L = 4$ m total grid size. Without performing any FTs, analytically show that the sampled signal has aliasing.

2. Show the sampling diagram for a point source with wavelength of a 1 μm

propagating a distance 100 km to a 2-m-diameter aperture.

3. Show the sampling diagram for a source with a wavelength of 0.5 μm and a diameter of 1 mm propagating a distance 2.0 m to a 2-m-diameter aperture.

4. Modify Listings 7.2 and B.5 to use a converging/diverging source of the form

$$U(x_1, y_1) = \text{rect}\left(\frac{x_1}{D_1}\right)\text{rect}\left(\frac{y_1}{D_1}\right)e^{i\frac{k}{2R}\left(x_1^2+y_1^2\right)}. \tag{7.62}$$

   (a) Rework the analytic solution for Fresnel diffraction by a square aperture given in Eq. (1.60) to include the diverging/converging wavefront in Eq. (7.62). Just a little algebraic manipulation obtains an analytic result similar to Eq. (1.60), but slightly more general to account for the diverging/converging source. See Ref. 5 for details on the derivation of Eq. (1.60).

   (b) Let $D_1 = 2$ mm, $D_2 = 4$ mm, $\Delta z = 0.1$ m, $\lambda = 1\ \mu$m, and $R = -0.2$ m (just like in the example, but with a converging source). In preparation for carrying out an angular-spectrum simulation, generate plots similar to Fig. 7.5 to show your careful method of picking values for δ_1, δ_2, and N.

   (c) Carry out the simulation, and produce plots of the $y_2 = 0$ slice of the amplitude and phase. Evaluate the analytic result you obtained in part (a) for the given parameters, and include the analytic result on those same plots.

5. Show diagrammatically that Eq. (7.60) means that the geometric beam is contained with a region of diameter D_2. Show the ray diagrams for diverging source fields and converging source fields that are focused in front of and behind the observation plane.

6. Show algebraically that constraint 4 is more restrictive than the combination of constraints 1 and 2.

Chapter 8
Relaxed Sampling Constraints with Partial Propagations

The sampling constraints for Fresnel propagation are strict. Particularly, the angular-spectrum method is best suited for propagating only short distances. The key problem is wrap-around, caused by aliasing. Several approaches to mitigating these effects have been proposed. Most of these approaches center around spatially attenuating and filtering the optical field. For example, Johnston and Lane describe a technique in which the free-space transfer function is filtered and the grid size is based on the bandwidth of the filter.[41] After this step, they set the sample interval based on avoiding aliasing of the quadratic phase factor just like in Sec. 7.3.2.

Johnston and Lane's choice of spatial-filter bandwidth works, but it is somewhat indirectly related to specific wrap-around effects. This book covers a more direct approach. For fixed D_1, δ_1, D_2, and δ_2, we must satisfy constraints 1, 3, and 4 from Ch. 7. Generally, Δz is fixed, too; it is just a part of the geometry that we wish to simulate. Often, the only free parameter is N, and for large Δz the constraints dictate large N. Sometimes the required N is prohibitively large, like $N > 4096$. Usually the culprit is constraint 4, which is only dependent on the propagation method, not the fixed propagation geometry. If constraint 4 is satisfied, it remains satisfied if we shorten Δz while holding N, δ_1, δ_2, and λ fixed. Consequently, this chapter develops a method of using multiple partial propagations with the angular-spectrum method to significantly relax constraint 4. To illustrate the propagation algorithm, we first begin with two partial propagations in Sec. 8.2 and then generalize to $n - 1$ partial propagations (n planes) in Sec. 8.3.

At first this may sound like a good solution, but multiple partial propagations are mathematically equivalent to a single full propagation. The extra partial propagations just take longer to execute. The key difficulty that we want to mitigate is wrap-around caused by aliasing. The variations in the free-space transfer function, given in Eq. (6.32), become increasingly rapid as Δz increases. Therefore, wrap-around effects creep into the center of the grid from the edge. With partial propagations, we can attenuate the field at the edges of the grid to suppress the wrap-around all along the path. This method allows us to increase the useable range of conditions for our simulation method or reduce the grid size at the cost of executing more

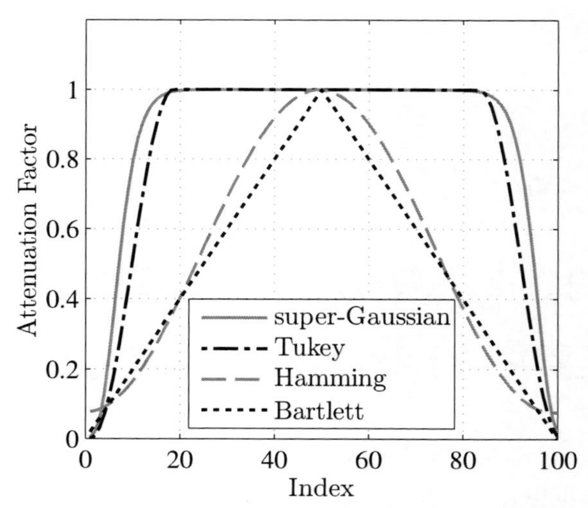

Figure 8.1 Examples of data windows. The super-Gaussian and Tukey windows are appropriate for optical simulations, while the Hamming and Bartlett windows are not. The super-Gaussian shown has $\sigma = 0.45L$ and $n = 16$, while the Tukey window shown has $\alpha = 0.65$.

propagations. In most cases, this shortens the simulation's execution time.

8.1 Absorbing Boundaries

Attenuating the field at the edge has the effect of absorbing energy that is spreading beyond the extent of the grid. The operation is to simply multiply the field by an attenuating factor at each partial-propagation plane. This is similar to the concept of data windowing, but we must be careful not to alter light in the central region of the grid. For this reason, the attenuating factor is very close to unity in the center of the grid and very close to zero at the edge. Common data windows, such as the Hamming and Bartlett windows, are not suited for this purpose. Examples of well suited attenuation factors are the super-Gaussian function defined by

$$g_{sg}(x, y) = \exp\left[-\left(\frac{r}{\sigma}\right)^n\right], \quad n > 2, \tag{8.1}$$

where $n > 2$ and the Tukey (or cosine-taper) window defined by

$$g_{ct}(x, y) = \begin{cases} 1 & r \geq \alpha L/2 \\ \frac{1}{2}\left\{1 + \cos\left[\pi \frac{r/L - \alpha N/2}{(1-\alpha)N/2}\right]\right\} & \alpha N/2 \leq r/L \leq N/2, \end{cases} \tag{8.2}$$

where $0 \leq \alpha \leq 1$ is a parameter that specifies the width of the tapered region. Large α values specify a broad unattenuated region in the center and narrow taper at the edges. These windows are shown in Fig. 8.1.

Absorbing boundaries have been used several times in the literature. For example, Flatté, *et al.* used a super-Gaussian with $n = 8$ to model a plane wave in their

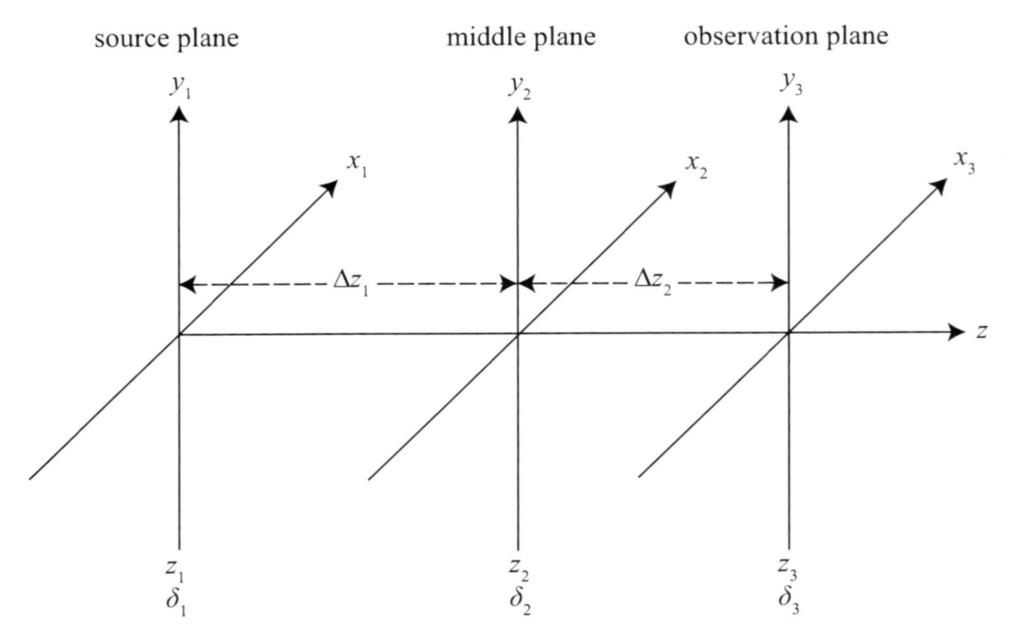

Figure 8.2 Coordinate systems for two partial propagations.

studies of turbulent propagation.[53] Later Rubio adopted the same type of super-Gaussian specifically as an absorbing boundary all along the propagation path.[33] It was used to contain the energy from a diverging spherical wave. The Tukey window was used by Frehlich in his studies of generating atmospheric phase screens.[56]

As an additional example of an absorbing boundary that is not a widely used window, Martin and Flatté used a Gaussian extinction coefficient in their simulations of propagation through atmospheric turbulence.[44] To do so, they added a deterministic imaginary component to their random atmospheric phase screens, thereby multiplying log-amplitude by a Gaussian factor at the edges of the grid. The extinction coefficient in the center of the grid was set to zero so that the field in the center was not attenuated.

8.2 Two Partial Propagations

In this subsection, we simply perform angular-spectrum propagation twice. The first propagation goes from the the source plane to the "middle" plane (somewhere between the source and observation planes, not necessarily half-way), and the second propagation goes from the middle plane to the observation plane. The absorbing boundary is applied in the middle plane after the first propagation. The geometry for two partial propagations is illustrated in Fig. 8.2. The symbols for this subsection are defined in Table 8.1.

Before we get into the simulation equations, we need to determine some mathematical relationships among the symbols in Table 8.1. Figure 8.3 shows the geometry of grid spacings. In the figure, A and B are grid points in the source plane, so

Table 8.1 Definition of symbols for performing two partial propagations.

symbol	meaning
$\mathbf{r}_1 = (x_1, y_1)$	source-plane coordinates
$\mathbf{r}_2 = (x_2, y_2)$	middle-plane coordinates
$\mathbf{r}_3 = (x_3, y_3)$	observation-plane coordinates
δ_1	grid spacing in source plane
δ_2	grid spacing in middle plane
δ_3	grid spacing in observation plane
$\mathbf{f}_1 = (f_{x1}, f_{y1})$	spatial frequency of source plane
$\mathbf{f}_2 = (f_{x2}, f_{y2})$	spatial frequency of middle plane
δ_{f1}	grid spacing in source-plane spatial frequency
δ_{f2}	grid spacing in middle-plane spatial frequency
$z_1 = 0$	location of source plane along the optical axis
z_2	location of middle plane along the optical axis
z_3	location of observation plane along the optical axis
Δz_1	distance between source plane and middle plane
Δz_2	distance between middle plane and observation plane
$\Delta z = \Delta z_1 + \Delta z_2$	distance between source plane and observation plane
$\alpha = \Delta z_1 / \Delta z$	fractional distance of first propagation
m	scaling factor from source plane to observation plane
m_1	scaling factor from source plane to middle plane
m_2	scaling factor from middle plane to observation plane

they are separated by a distance δ_1, consistent with Table 8.1. Points C and D are grid points in the middle plane, so according to Table 8.1, they are separated by a distance δ_2. Finally, E and F are grid points in the observation plane, so they are separated by a distance δ_3. Triangles $\triangle BDH$ and $\triangle BFG$ share a vertex, so they are similar triangles. Therefore, their side lengths are related by

$$\frac{\overline{DH}}{\overline{BH}} = \frac{\overline{FG}}{\overline{BG}}. \tag{8.3}$$

The length of segment $\overline{FG}$ is $(\delta_3 - \delta_1)/2$, and the length of segment $\overline{DH}$ is $(\delta_2 - \delta_1)/2$. The length of segment $\overline{BH}$ is Δz_1, and the length of segment $\overline{BG}$ is $\Delta z = \Delta z_1 + \Delta z_2$. With this knowledge, Eq. (8.3) becomes

$$\frac{\delta_2 - \delta_1}{2\,\Delta z_1} = \frac{\delta_3 - \delta_1}{2\,\Delta z} \tag{8.4}$$

$$\delta_2\,\Delta z - \delta_1\,\Delta z = \delta_3\,\Delta z_1 - \delta_1\,\Delta z_1 \tag{8.5}$$

$$\delta_2 = \delta_1 + \frac{\delta_3\,\Delta z_1 - \delta_1\,\Delta z_1}{\Delta z} \tag{8.6}$$

$$\delta_2 = \delta_1 + \alpha\,\delta_3 - \alpha\,\delta_1 \tag{8.7}$$

$$\delta_2 = (1 - \alpha)\,\delta_1 + \alpha\,\delta_3. \tag{8.8}$$

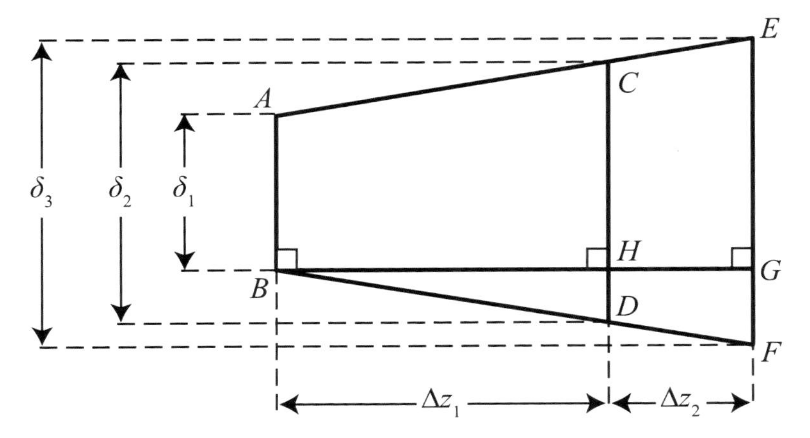

Figure 8.3 Grid spacings for partial propagations.

With these basic relationships among the propagation parameters now known, we can proceed with writing down the equation for performing two successive propagations. When propagating a distance Δz_1 to the middle plane and then propagating a distance Δz_2, the observation-plane field $U(\mathbf{r}_3)$ is given by

$$\begin{aligned} U(\mathbf{r}_3) = &\ \mathcal{Q}\left[\frac{m_2-1}{m_2\,\Delta z_2}, \mathbf{r}_3\right] \mathcal{F}^{-1}\left[\mathbf{f}_2, \frac{\mathbf{r}_3}{m_2}\right] \\ &\times \mathcal{Q}_2\left[-\frac{\Delta z_2}{m_2}, \mathbf{f}_2\right] \mathcal{F}\left[\mathbf{r}_2, \mathbf{f}_2\right] \mathcal{Q}\left[\frac{1-m_2}{\Delta z_2}, \mathbf{r}_2\right] \frac{1}{m_2} \\ &\times \mathcal{A}\left[\mathbf{r}_2\right] \mathcal{Q}\left[\frac{m_1-1}{m_1\,\Delta z_1}, \mathbf{r}_2\right] \mathcal{F}^{-1}\left[\mathbf{f}_1, \frac{\mathbf{r}_2}{m_1}\right] \mathcal{Q}_2\left[-\frac{\Delta z_1}{m_1}, \mathbf{f}_1\right] \\ &\times \mathcal{F}\left[\mathbf{r}_1, \mathbf{f}_1\right] \mathcal{Q}\left[\frac{1-m_1}{\Delta z_1}, \mathbf{r}_1\right] \frac{1}{m_1} \left\{U(\mathbf{r}_1)\right\}, \end{aligned} \tag{8.9}$$

where $\mathcal{A}[\mathbf{r}_2]$ is the operator corresponding to the absorbing boundary that is applied to the field in plane 2 (super-Gaussian, Tukey, or similar). The effect of this operator is to multiply the field by a function which reduces the field's amplitude near the edge of the grid.

The quadratic phase factors and the absorbing boundary all commute with each other because they are just multiplicative factors. This may allow us to combine the two middle-plane quadratic phase factors, thus eliminating a step and gaining a little computational efficiency. The product

$$\mathcal{Q}\left[\frac{1-m_2}{\Delta z_2}, \mathbf{r}_2\right] \mathcal{Q}\left[\frac{m_1-1}{m_1\,\Delta z_1}, \mathbf{r}_2\right]$$

can be simplified. To do so, we seek a relationship between the arguments $(1-m_2)$ $/\Delta z_2$ and $(m_1-1)/(m_1\,\Delta z_1)$. Let us revisit Eq. (8.5) and work the factors m_1 and m_2 into the equation

$$\delta_2\,\Delta z - \delta_1\,\Delta z = \delta_3\,\Delta z_1 - \delta_1\,\Delta z_1 \tag{8.10}$$

$$\delta_2\,\Delta z_1 + \delta_2\,\Delta z_2 - \delta_1\,\Delta z_1 - \delta_1\,\Delta z_2 = \delta_3\,\Delta z_1 - \delta_1\,\Delta z_1 \tag{8.11}$$

$$\delta_3\Delta z_1 - \delta_2\Delta z_1 = \delta_2\Delta z_2 - \delta_1\Delta z_2 \tag{8.12}$$

$$\frac{\delta_3 - \delta_2}{\Delta z_2} = \frac{\delta_2 - \delta_1}{\Delta z_1} \tag{8.13}$$

$$\frac{\delta_3 - \delta_2}{\delta_2\Delta z_2} = \frac{\delta_2 - \delta_1}{\delta_2\Delta z_1} \tag{8.14}$$

$$\frac{m_2 - 1}{\Delta z_2} = \frac{m_1 - 1}{m_1\,\Delta z_1} \tag{8.15}$$

Therefore, the quadratic phase factors become

$$\mathcal{Q}\left[\frac{1-m_2}{\Delta z_2}, \mathbf{r}_2\right]\mathcal{Q}\left[\frac{m_1-1}{m_1\,\Delta z_1}, \mathbf{r}_2\right] = \mathcal{Q}\left[-\frac{m_1-1}{m_1\,\Delta z_1}, \mathbf{r}_2\right]\mathcal{Q}\left[\frac{m_1-1}{m_1\,\Delta z_1}, \mathbf{r}_2\right] = 1.$$

With this simplification, Eq. (8.9) becomes

$$U\left(\mathbf{r}_3\right) = \mathcal{Q}\left[\frac{m_2-1}{m_2\,\Delta z_2}, \mathbf{r}_3\right]\mathcal{F}^{-1}\left[\mathbf{f}_2, \frac{\mathbf{r}_3}{m_2}\right]\mathcal{Q}_2\left[-\frac{\Delta z_2}{m_2}, \mathbf{f}_2\right]\mathcal{F}\left[\mathbf{r}_2, \mathbf{f}_2\right]\frac{1}{m_2}$$
$$\times\mathcal{A}\left[\mathbf{r}_2\right]\mathcal{F}^{-1}\left[\mathbf{f}_1, \frac{\mathbf{r}_2}{m_1}\right]\mathcal{Q}_2\left[-\frac{\Delta z_1}{m_1}, \mathbf{f}_1\right]\mathcal{F}\left[\mathbf{r}_1, \mathbf{f}_1\right]\mathcal{Q}\left[\frac{1-m_1}{\Delta z_1}, \mathbf{r}_1\right]\frac{1}{m_1}\left\{U\left(\mathbf{r}_1\right)\right\}. \tag{8.16}$$

This specific result is not implemented in any simulation, but it helps establish a pattern for use with an arbitrary number of partial propagations.

8.3 Arbitrary Number of Partial Propagations

To get a useful result from the previous section, we must generalize it to an arbitrary number of partial propagations. First, let us write the table of propagation and simulation parameters more generally. These parameters are given in Table 8.2 for n propagation planes and $n-1$ partial propagations. As examples, the quantities for the first propagation are given in Table 8.3, and the quantities for the second propagation are given in Table 8.4.

Let us reorder (when possible) and group factors in Eq. (8.16) so that

$$\begin{aligned} U\left(\mathbf{r}_3\right) &= \mathcal{Q}\left[\frac{m_2-1}{m_2\,\Delta z_2}, \mathbf{r}_3\right]\left\{\mathcal{F}^{-1}\left[\mathbf{f}_2, \frac{\mathbf{r}_3}{m_2}\right]\mathcal{Q}_2\left[-\frac{\Delta z_2}{m_2}, \mathbf{f}_2\right]\mathcal{F}\left[\mathbf{r}_2, \mathbf{f}_2\right]\frac{1}{m_2}\right\} \\ &\times\left\{\mathcal{A}\left[\mathbf{r}_2\right]\mathcal{F}^{-1}\left[\mathbf{f}_1, \frac{\mathbf{r}_2}{m_1}\right]\mathcal{Q}_2\left[-\frac{\Delta z_1}{m_1}, \mathbf{f}_1\right]\mathcal{F}\left[\mathbf{r}_1, \mathbf{f}_1\right]\frac{1}{m_1}\right\} \\ &\times\left\{\mathcal{Q}\left[\frac{1-m_1}{\Delta z_1}, \mathbf{r}_1\right]U\left(\mathbf{r}_1\right)\right\}. \end{aligned} \tag{8.17}$$

Now, it is clear what operations are repeated for each partial propagation, so it is straightforward to generalize this to $n-1$ partial propagations:

$$U\left(\mathbf{r}_n\right) = \mathcal{Q}\left[\frac{m_{n-1}-1}{m_{n-1}\,\Delta z_{n-1}}, \mathbf{r}_n\right]$$

$$\times \prod_{i=1}^{n-1} \left\{ \mathcal{A}\left[\mathbf{r}_{i+1}\right] \mathcal{F}^{-1}\left[\mathbf{f}_i, \frac{\mathbf{r}_{i+1}}{m_i}\right] \mathcal{Q}_2\left[-\frac{\Delta z_i}{m_i}, \mathbf{f}_i\right] \mathcal{F}\left[\mathbf{r}_i, \mathbf{f}_i\right] \frac{1}{m_i} \right\}$$
$$\times \left\{ \mathcal{Q}\left[\frac{1-m_1}{\Delta z_1}, \mathbf{r}_1\right] U\left(\mathbf{r}_1\right) \right\}. \tag{8.18}$$

Listing 8.1 shows code for evaluating the Fresnel diffraction integral in MATLAB using an arbitrary number of partial propagations with the angular-spectrum method. In the listing, the inputs are

`Uin` : $U(\mathbf{r}_1)$, the optical field in the source plane $[(\mathrm{W/m}^2)^{1/2}]$,

`wvl` : λ, the optical wavelength (m),

`delta1` : δ_1, grid spacing in the source plane (m),

`deltan` : δ_n, grid spacing in the observation plane (m),

`z` : an array containing the values of z_i for $i = 2, 3, \ldots n$ (m).

The outputs are

`xn` : x coordinates in the observation plane (m),

`yn` : y coordinates in the observation plane (m),

`Uout` : $U(\mathbf{r}_n)$, optical field values in the observation plane $[(\mathrm{W/m}^2)^{1/2}]$.

After the sampling is discussed in the next section, an example simulation is presented to illustrate the accuracy of this method.

8.4 Sampling for Multiple Partial Propagations

With an arbitrary number of planes and repeated partial propagations, the sampling constraints must be re-examined. Chapter 7 discusses proper sampling for one complete propagation in detail. It includes a set of four inequalities that must be satisfied when choosing grid spacings and the number of grid points. The first two inequalities are based on the propagation geometry, not the propagation method, so when using multiple partial propagations, they remain unchanged. However, the last two inequalities prevent aliasing of two quadratic phase factors, which depend on grid spacings and propagation distance. The grid spacings and propagation distances can change for every partial propagation, so we need to modify our approach.

Recall that constraint 3 is based on avoiding aliasing of the quadratic phase factor inside the FT of the angular-spectrum method. The same concept applies here. Again we assume a spherical source wavefront with radius R so that the combined phase of the source field and the quadratic phase factor is

$$\phi = \frac{k}{2}\left(\frac{1-m_1}{\Delta z_1} + \frac{1}{R}\right)\left|\mathbf{r}_1\right|^2. \tag{8.19}$$

Table 8.2 Definition of symbols for performing an arbitrary number of partial propagations.

quantity	description
n	number of planes
$n-1$	number of propagations
for the ith propagation	
$\Delta z_i = z_{i+1} - z_i$	propagation distance from plane i to plane $i+1$
$\alpha_i = z_i / \Delta z$	fractional distance from plane 1 to plane $i+1$
$m_i = \delta_{i+1} / \delta_i$	scaling factor from plane i to plane $i+1$
source plane has	
$\mathbf{r}_i = (x_i, y_i)$	coordinates
$\delta_i = (1-\alpha_i)\,\delta_1 + \alpha_i \delta_n$	grid spacing in the ith plane
$\mathbf{f}_i = (f_{xi}, f_{yi})$	spatial-frequency coordinates
$\delta_{fi} = 1/(N\delta_i)$	grid spacing in spatial-frequency domain
observation plane has	
$\mathbf{r}_{i+1} = (x_{i+1}, y_{i+1})$	coordinates
$\delta_{i+1} = (1-\alpha_{i+1})\,\delta_1 + \alpha_{i+1}\delta_n$	grid spacing

Table 8.3 Symbols for performing the first of an arbitrary number of partial propagations.

symbol	meaning
for the 1st propagation	
$\Delta z_1 = z_2 - z_1$	propagation distance from plane 1 to plane 2
$\alpha_1 = z_1/\Delta z = 0$	fractional distance from plane 1 to plane 1
$\alpha_2 = z_2/\Delta z$	fractional distance from plane 1 to plane 2
$m_1 = \delta_2/\delta_1$	scaling factor from plane 1 to plane 2
source has	
$\mathbf{r}_1 = (x_1, y_1)$	coordinates
δ_1	grid spacing in the 1st plane
$\mathbf{f}_1 = (f_{x1}, f_{y1})$	spatial-frequency coordinates
δ_{f1}	grid spacing in spatial-frequency domain
observation plane has	
$\mathbf{r}_2 = (x_2, y_2)$	coordinates
δ_2	grid spacing

Table 8.4 Symbols for performing the second of an arbitrary number of partial propagations.

symbol	meaning
for the 2nd propagation	
$\Delta z_2 = z_3 - z_2$	propagation distance from plane 2 to plane 3
$\alpha_2 = z_2/\Delta z$	fractional distance from plane 1 to plane 2
$\alpha_3 = z_3/\Delta z$	fractional distance from plane 1 to plane 3
$m_2 = \delta_3/\delta_2$	scaling factor from plane 2 to plane 3
source has	
$\mathbf{r}_2 = (x_2, y_2)$	coordinates
δ_2	grid spacing in the 2nd plane
$\mathbf{f}_2 = (f_{x2}, f_{y2})$	spatial-frequency coordinates
δ_{f2}	grid spacing in spatial-frequency domain
observation plane has	
$\mathbf{r}_3 = (x_3, y_3)$	coordinates
δ_3	grid spacing

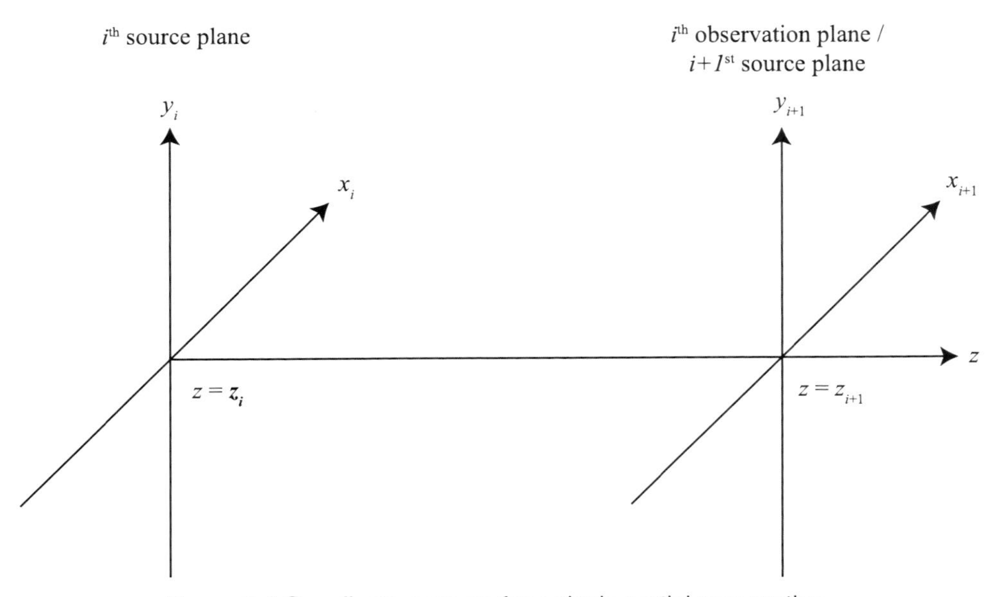

Figure 8.4 Coordinate systems for a single partial propagation.

Listing 8.1 Code for evaluating the Fresnel diffraction integral in MATLAB using an arbitrary number of partial propagations with the angular-spectrum method.

```
1  function [xn yn Uout] = ang_spec_multi_prop_vac ...
2      (Uin, wvl, delta1, deltan, z)
3  % function [xn yn Uout] = ang_spec_multi_prop_vac ...
4  %     (Uin, wvl, delta1, deltan, z)
5
6      N = size(Uin, 1);   % number of grid points
7      [nx ny] = meshgrid((-N/2 : 1 : N/2 - 1));
8      k = 2*pi/wvl;    % optical wavevector
9      % super-Gaussian absorbing boundary
10     nsq = nx.^2 + ny.^2;
11     w = 0.47*N;
12     sg = exp(-nsq.^8/w^16); clear('nsq', 'w');
13
14     z = [0 z];  % propagation plane locations
15     n = length(z);
16     % propagation distances
17     Delta_z = z(2:n) - z(1:n-1);
18     % grid spacings
19     alpha = z / z(n);
20     delta = (1-alpha) * delta1 + alpha * deltan;
21     m = delta(2:n) ./ delta(1:n-1);
22     x1 = nx * delta(1);
23     y1 = ny * delta(1);
24     r1sq = x1.^2 + y1.^2;
25
26     Q1 = exp(i*k/2*(1-m(1))/Delta_z(1)*r1sq);
27     Uin = Uin .* Q1;
28     for idx = 1 : n-1
29         % spatial frequencies (of i^th plane)
30         deltaf = 1 / (N*delta(idx));
31         fX = nx * deltaf;
32         fY = ny * deltaf;
33         fsq = fX.^2 + fY.^2;
34         Z = Delta_z(idx);   % propagation distance
35         % quadratic phase factor
36         Q2 = exp(-i*pi^2*2*Z/m(idx)/k*fsq);
37         % compute the propagated field
38         Uin = sg .* ift2(Q2 ...
39             .* ft2(Uin / m(idx), delta(idx)), deltaf);
40     end
41     % observation-plane coordinates
42     xn = nx * delta(n);
43     yn = ny * delta(n);
44     rnsq = xn.^2 + yn.^2;
45     Q3 = exp(i*k/2*(m(n-1)-1)/(m(n-1)*Z)*rnsq);
46     Uout = Q3 .* Uin;
```

At first, this constraint looks confusing because it depends on Δz_1, and we cannot determine Δz_1 until the rest of the sampling analysis is complete! Nonetheless, we carry on with the analysis. It proceeds just like in Eqs. (7.48)–(7.53) to yield

$$\left(1+\frac{\Delta z_1}{R}\right)\delta_1-\frac{\lambda\Delta z_1}{D_1}\leq\delta_2\leq\left(1+\frac{\Delta z_1}{R}\right)\delta_1+\frac{\lambda\Delta z_1}{D_1}. \tag{8.20}$$

Now, we substitute in for δ_2 and Δz_1 to get

$$\left(1+\frac{\alpha_2\Delta z}{R}\right)\delta_1-\frac{\lambda\alpha_2\Delta z}{D_1}\leq(1-\alpha_2)\,\delta_1+\alpha_2\delta_n\leq\left(1+\frac{\alpha_2\Delta z}{R}\right)\delta_1+\frac{\lambda\alpha_2\Delta z}{D_1}. \tag{8.21}$$

After multiplying everything out and eliminating common terms, we are left with

$$\left(1+\frac{\Delta z}{R}\right)\delta_1-\frac{\lambda\Delta z}{D_1}\leq\delta_n\leq\left(1+\frac{\Delta z}{R}\right)\delta_1+\frac{\lambda\Delta z}{D_1}. \tag{8.22}$$

This is identical to Eq. (7.53), which has no dependence on quantities related to partial-propagation planes, like δ_2 and Δz_1!

Now, constraint 4 is the only one left to modify, and we must find a way to relate it to n. Hopefully, it is related in such a way that n partial propagations relaxes this constraint. For the the ith partial propagation, it is given by

$$N\geq\frac{\lambda\Delta z_i}{\delta_i\delta_{i+1}}. \tag{8.23}$$

This makes a very complicated parameter space. To simplify, we can write all δ_i in terms of δ_1 and δ_n. However, that just exchanges δ_i for α_i, which depends on z_i. There is just no way to reduce the dimensions of the parameter space for this constraint. Rather than trying to satisfy all n constraints implied by Eq. (8.23), we only need to satisfy the case for which the right-hand side is a maximum. However, that requires prior knowledge of all the Δz_i and δ_i, which is what we are trying to determine!

Obviously, a new approach is necessary. Let us write down the inequalities again and regroup

1. $\delta_n\leq\frac{\lambda\Delta z-D_2\delta_1}{D_1}$,
2. $N\geq\frac{D_1}{2\delta_1}+\frac{D_2}{2\delta_n}+\frac{\lambda\Delta z}{2\delta_1\delta_n}$,
3. $\left(1+\frac{\Delta z}{R}\right)\delta_1-\frac{\lambda\Delta z}{D_1}\leq\delta_n\leq\left(1+\frac{\Delta z}{R}\right)\delta_1+\frac{\lambda\Delta z}{D_1}$,
4. $N\geq\frac{\lambda\Delta z_i}{\delta_i\delta_{i+1}}$.

Examining the inequalities, we can see that it is possible to use the first three inequalities to choose values of N, δ_1, and δ_n. Then, we can find a way to satisfy the fourth constraint.

Depending on whether we are using expanding or contracting propagation grids, either δ_1 or δ_n is smaller than all other δ_i. For a given value of Δz_i, picking the smaller of δ_1 and δ_n to replace δ_i and δ_{i+1} in the fourth inequality gives us a single constraint that N must satisfy. However, N is already chosen using the first two constraints, and the limit on Δz_i remains unknown, so we must rewrite the inequality as a constraint on Δz_i so that

$$\Delta z_i \leq \frac{\min\left(\delta_1, \delta_n\right)^2 N}{\lambda}. \tag{8.24}$$

The right-hand side is the maximum possible partial-propagation distance Δz_{max} that can be used. Therefore, we must use at least $n = \text{ceil}\left(\Delta z / \Delta z_{max}\right) + 1$ partial propagations (where ceil is the "ceiling" function; it produces the smallest integer value that is greater than or equal to its argument).

Finally, with this new view of the fourth inequality, the method of choosing propagation-grid parameters is clear:

1. First, pick N, δ_1, and δ_n based on the first two inequalities.
2. Then, use a slightly adjusted version of the fourth inequality [Eq. (8.24)] to determine the maximum partial-propagation distance and the minimum number of partial propagations $n - 1$ together.
3. One can always use more partial propagations; shorter partial-propagation distances still satisfy the fourth inequality.

We close this chapter with an example of using this method to achieve accurate results within the observation-plane region of interest. In this example, we want to simulate propagation of a uniform-amplitude plane wave ($R = \infty$) departing a square aperture in the source plane. The aperture has $D_1 = 2$ mm across each side. The optical wavelength is $\lambda = 1\,\mu$m, and the sensor is in the observation plane located $\Delta z = 2$ m from the source plane. Figure 8.5 shows a contour plot of constraint 2 with a plot of constraint 1 overlayed. Often, it is helpful to have a certain number of grid points across the source aperture and the observation-plane region of interest. For this example, we choose to have at least 30 grid points across D_1 and D_2. This choice gives $\delta_1 \leq 66.7\,\mu$m and $\delta_n \leq 133\,\mu$m. According to the contour plot, at least $N = 2^7 = 128$ grid points are required. To conclude the sampling analysis, we apply constraint 4 with $\delta_1 = 66.7\,\mu$m, $\delta_n = 133\,\mu$m, and $N = 128$. This gives

$$\Delta z_{max} = \frac{\min\left(\delta_1, \delta_n\right)^2 N}{\lambda} = \frac{\left(66.7\,\mu\text{m}\right)^2\, 128}{1\,\mu\text{m}} = 0.567\,\text{m}. \tag{8.25}$$

Then, we need to perform at least $n = \text{ceil}\left(2\,\text{m}/0.567\,\text{m}\right) + 1 = 5$ partial propagations. Listing 8.2 gives the MATLAB code used to simulate the propagation for this

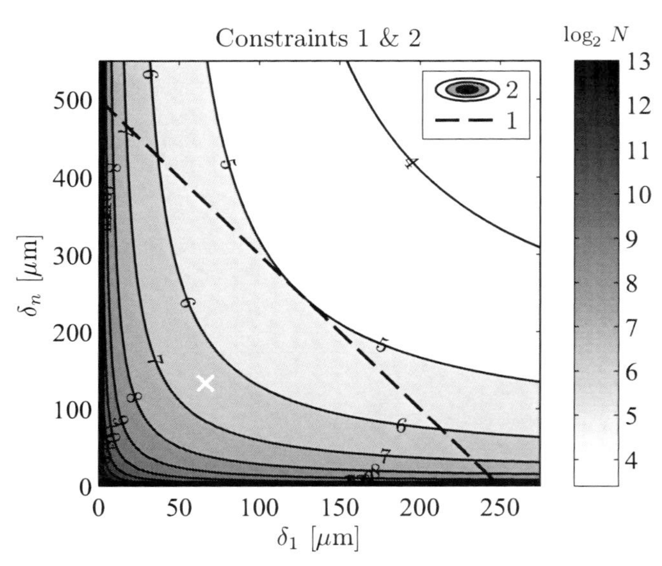

Figure 8.5 Analysis of sampling constraints. The white x marks grid spacings that correspond to 30 grid points across the source- and observation-plane apertures.

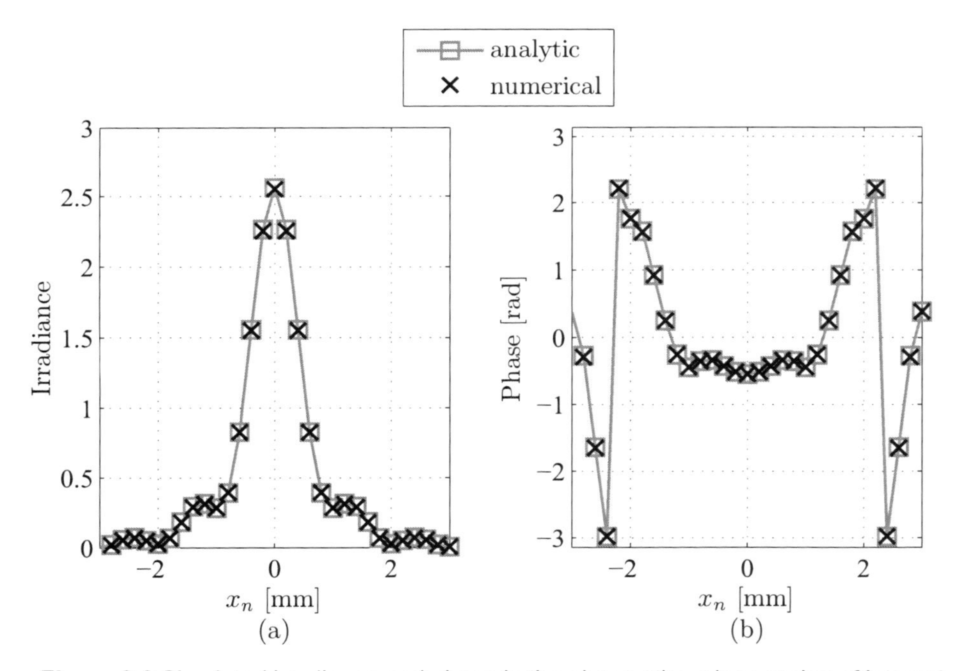

Figure 8.6 Simulated irradiance and phase in the observation-plane region of interest.

example. Figure 8.6 shows the simulated irradiance and phase in the observation-plane region of interest. As usual, the simulation result matches the theoretical expectation closely in the observation-plane region of interest.

Listing 8.2 Example of evaluating the Fresnel diffraction integral in MATLAB using the angular-spectrum method with several partial propagations.

```
1  % example_square_prop_ang_spec_multi.m
2
3  D1 = 2e-3;    % diameter of the source aperture [m]
4  D2 = 6e-3;    % diameter of the observation aperture [m]
5  wvl = 1e-6;  % optical wavelength [m]
6  k = 2*pi / wvl; % optical wavenumber [rad/m]
7  z = 2;       % propagation distance [m]
8  delta1 = D1/30; % source-plane grid spacing [m]
9  deltan = D2/30; % observation-plane grid spacing [m]
10 N = 128;        % number of grid points
11 n = 5;          % number of partial propagations
12 % switch from total distance to individual distances
13 z = (1:n) * z / n;
14 % source-plane coordinates
15 [x1 y1] = meshgrid((-N/2 : N/2-1) * delta1);
16 ap = rect(x1/D1) .* rect(y1/D1);     % source aperture
17 [x2 y2 Uout] = ...
18     ang_spec_multi_prop_vac(ap, wvl, delta1, deltan, z);
19
20 % analytic result for y2=0 slice
21 Dz = z(end); % switch back to total distance
22 Uout_an ...
23     = fresnel_prop_square_ap(x2(N/2+1,:), 0, D1, wvl, Dz);
```

8.5 Problems

1. Consider the signal

$$g(x) = \exp\left(i\pi a^2 x^2\right) \tag{8.26}$$

with $a = 4$ sampled on a grid with $N = 128$ points and $L = 4$ m total grid size. Compute both the analytic and discrete FT of this signal. Next, premultiply the signal by a super-Gaussian absorbing boundary function with $n = 16$ and $\sigma = 0.25L$ and compute the DFT again. Plot the imaginary and real parts of the two DFT results (with and without the absorbing boundary) and compare against the analytic FT.

2. Fill in the missing steps between Eqs. (8.20) and (8.22) to show that constraint 3 is identical for any number of partial propagations.

3. Show the sampling diagram for a point source with wavelength 1 μm propagating 100 km to a telescope with a 2-m-diameter aperture. How does this compare to the case when there is only one propagation?

4. Simulate propagation of a uniform-amplitude plane wave from an annular aperture to a target plane with the source beam focused onto the target. Let the annular aperture have an outer diameter of 1.5 m and an inner diameter of 0.5 m. Let the optical wavelength be 1.3 μm. Place the target in the observation plane 100 km away from the source plane.

    (a) Show a detailed sampling analysis similar to that shown in Fig. 8.5. Be sure to describe your analysis of determining the number of partial propagations to use.

    (b) After completing the simulation, show plots of the $y_n = 0$ slice of the observation-plane irradiance and phase. Include the analytic and simulation results on the same plot.

Chapter 9

Propagation through Atmospheric Turbulence

Up to this point, the propagation algorithms have been designed to simulate propagation through vacuum and through simple optical systems that can be described by ray matrices. There are several other more complicated and useful applications of the split-step beam propagation method. These include sources with partial temporal and spatial coherence, coherent propagation through deterministic structures like fibers and integrated optical devices, and propagation through random media like atmospheric turbulence. This chapter focuses on coherent propagation through turbulence, and the method is shown to be very closely related to propagation through vacuum.

Earth's atmosphere is a medium whose refractive index is nearly unity. This allows us to make only a slight modification to our vacuum-propagation techniques from Ch. 8 to simulate propagation through the atmosphere. Unfortunately, the atmosphere's refractive index randomly evolves over space and time. This effect causes light to be randomly distorted as it propagates. As a result, optical systems that rely on light propagating through the atmosphere must overcome a great challenge. For example, astronomers have observed for centuries that atmospheric turbulence limits the resolution of their telescopes. This is why observatories are built on mountain tops; the location minimizes the turbulent path distance through which the light must propagate.

To simulate atmospheric propagation, we first develop the simulation algorithm, and then we discuss atmospheric turbulence and how to model its refractive properties. Finally, we discuss setting up an atmospheric simulation, proper sampling with due consideration to the effects of the atmosphere, and verifying that the output is consistent with analytic theory.

9.1 Split-Step Beam Propagation Method

Simulating propagation through non-vacuum media is accomplished through the split-step beam propagation method.[40,57–59] This method is useful for simulating propagation through many types of materials: inhomogeneous, anisotropic, and nonlinear. In this chapter, the discussion is restricted to the atmosphere, which is a linear, isotropic material with inhomogeneous refractive index n, i.e., $n = n(x, y, z)$. When $\delta n = n - 1$ is small, it can be shown that the field in the $i + 1^{\text{st}}$ plane is[59]

$$U(\mathbf{r}_{i+1}) \simeq \mathcal{R}\left[\frac{\Delta z_i}{2}, \mathbf{r}_i, \widetilde{\mathbf{r}}_{i+1}\right] \mathcal{T}[z_i, z_{i+1}] \mathcal{R}\left[\frac{\Delta z_i}{2}, \mathbf{r}_i, \widetilde{\mathbf{r}}_{i+1}\right] \{U(\mathbf{r}_i)\}, \quad (9.1)$$

where $\mathcal{T}[z_i, z_{i+1}]$ is an operator representing the accumulation of phase and $\widetilde{\mathbf{r}}_{i+1}$ is a coordinate in a plane half-way between the ith and $i + 1^{\text{st}}$ planes. It is given by

$$\mathcal{T}[z_i, z_{i+1}] = \exp[-i\phi(\mathbf{r}_{i+1})], \quad (9.2)$$

where the accumulated phase is $\phi(\mathbf{r}_i) = k\int_{z_i}^{z_{i+1}} \delta n(\mathbf{r}_i)\, dz$. Equation (9.1) indicates that we can separate propagation through a medium into two effects: diffraction and refraction. Free-space diffraction is represented by the operator $\mathcal{R}$, while refraction is represented by the operator $\mathcal{T}$. This method is commonly used to simulate propagation though atmospheric turbulence. In fact, it is used to emulate propagation through turbulence in optics laboratories, too.[60,61] The method is to alternate steps of partial vacuum propagation with interaction between the light and the material.[32,43,44]

Writing this algorithm concretely, there is a slight modification to the vacuum propagation algorithm from Eq. (8.18), given by

$$\begin{aligned} U(\mathbf{r}_n) = {} & \mathcal{Q}\left[\frac{m_{n-1} - 1}{m_{n-1}\Delta z_{n-1}}, \mathbf{r}_n\right] \\ & \times \prod_{i=1}^{n-1} \left\{\mathcal{T}[z_i, z_{i+1}] \mathcal{F}^{-1}\left[\mathbf{f}_i, \frac{\mathbf{r}_{i+1}}{m_i}\right] \mathcal{Q}_2\left[-\frac{\Delta z_i}{m_i}, \mathbf{f}_i\right] \mathcal{F}[\mathbf{r}_i, \mathbf{f}_i] \frac{1}{m_i}\right\} \\ & \times \left\{\mathcal{Q}\left[\frac{1 - m_1}{\Delta z_1}, \mathbf{r}_1\right] \mathcal{T}[z_1, z_2] U(\mathbf{r}_1)\right\}. \end{aligned} \quad (9.3)$$

Recall that there are $n - 1$ propagations and n planes with interaction in each plane. MATLAB code for this algorithm is given in the `ang_spec_multi_prop` function, provided in Listing 9.1. Note that it can be used for vacuum propagation if $\mathcal{T} = 1$ at every step. Example of usage of the `ang_spec_multi_prop` function is given in Sec. 9.5.4 after a discussion of turbulence and how to generate realizations of $\mathcal{T}$.

9.2 Refractive Properties of Atmospheric Turbulence

In this section, the basic theory of atmospheric turbulence is presented. It begins with the original analysis of turbulent flow by Kolmogorov, which eventually led to statistical models of the refractive-index variation.[62] Then, perturbation theory is used with the model to solve Maxwell's equations to obtain useful statistical

Listing 9.1 Code for evaluating the Fresnel diffraction integral in MATLAB through a weakly refractive medium using the angular-spectrum method.

```
1  function [xn yn Uout] = ang_spec_multi_prop ...
2      (Uin, wvl, delta1, deltan, z, t)
3  % function [xn yn Uout] = ang_spec_multi_prop ...
4  %     (Uin, wvl, delta1, deltan, z, t)
5
6      N = size(Uin, 1);   % number of grid points
7      [nx ny] = meshgrid((-N/2 : 1 : N/2 - 1));
8      k = 2*pi/wvl;    % optical wavevector
9      % super-Gaussian absorbing boundary
10     nsq = nx.^2 + ny.^2;
11     w = 0.47*N;
12     sg = exp(-nsq.^8/w^16); clear('nsq', 'w');
13
14     z = [0 z];  % propagation plane locations
15     n = length(z);
16     % propagation distances
17     Delta_z = z(2:n) - z(1:n-1);
18     % grid spacings
19     alpha = z / z(n);
20     delta = (1-alpha) * delta1 + alpha * deltan;
21     m = delta(2:n) ./ delta(1:n-1);
22     x1 = nx * delta(1);
23     y1 = ny * delta(1);
24     r1sq = x1.^2 + y1.^2;
25     Q1 = exp(i*k/2*(1-m(1))/Delta_z(1)*r1sq);
26     Uin = Uin .* Q1 .* t(:,:,1);
27     for idx = 1 : n-1
28         % spatial frequencies (of i^th plane)
29         deltaf = 1 / (N*delta(idx));
30         fX = nx * deltaf;
31         fY = ny * deltaf;
32         fsq = fX.^2 + fY.^2;
33         Z = Delta_z(idx);    % propagation distance
34         % quadratic phase factor
35         Q2 = exp(-i*pi^2*2*Z/m(idx)/k*fsq);
36         % compute the propagated field
37         Uin = sg .* t(:,:,idx+1) ...
38             .* ift2(Q2 ...
39             .* ft2(Uin / m(idx), delta(idx)), deltaf);
40     end
41     % observation-plane coordinates
42     xn = nx * delta(n);
43     yn = ny * delta(n);
44     rnsq = xn.^2 + yn.^2;
45     Q3 = exp(i*k/2*(m(n-1)-1)/(m(n-1)*Z)*rnsq);
46     Uout = Q3 .* Uin;
```

properties of the observation-plane optical field. The variances, correlations, and spectral densities of properties like log-amplitude, phase, and irradiance are used for two primary purposes in conjunction with the simulations. The first use is to produce random draws of the interaction factor for the split-step beam propagation method, which is done in Sec. 9.3. Then, after simulating propagation through the turbulent medium, the observation-plane fields are processed to determine their statistical properties and compare them against theory in Sec. 9.5.5. This provides confirmation that the simulation is producing accurate results.

9.2.1 Kolmogorov theory of turbulence

Turbulence in Earth's atmosphere is caused by random variations in temperature and convective air motion, which alter the air's refractive index, both spatially and temporally. As optical waves propagate through the atmosphere, the waves are distorted by these fluctuations in refractive index. This distortion of light has frustrated astronomers for centuries because it degrades their images of celestial objects. To overcome this distortion, they needed an accurate physical model of turbulence and its effects on optical-wave propagation. Since turbulence affects all optical systems that rely on propagating light through long atmospheric paths, like laser communication systems and laser weapons, optical physicists and communications engineers have begun to address this problem more recently.

Over the last hundred years, modeling the effects of turbulence on optical propagation has received much attention. Much has been written on various theories and experimental verification thereof. The focus on statistical modeling has produced several useful theories. In these theories, it is necessary to resort to statistical analyses, because it is impossible to exactly describe the refractive index for all positions in space and all time. There are too many random behaviors and variables to account for in a closed-form solution. The most widely accepted theory of turbulent flow, due to its consistent agreement with observation, was first put forward by A. N. Kolmogorov.[62] Later, Obukhov[63] and independently Corrsin[64] adapted Kolmogorov's model to temperature fluctuations. Then, the theory of turbulent temperature fluctuations could be directly related to refractive-index fluctuations. This model is the basis for all contemporary theories of turbulence.[65]

Differential heating and cooling of Earth by sunlight and the diurnal cycle cause large-scale variations in the temperature of air. This process consequently creates wind. As air moves, it transitions from laminar flow to turbulent flow. In laminar flow, the velocity characteristics are uniform or at least change in a regular fashion. In turbulent flow, air of different temperatures mixes, so the velocity field is no longer uniform, and it acquires randomly distributed pockets of air, called turbulent eddies. These eddies have varying characteristic sizes and temperatures. Since the density of air, and thus its refractive index, depends on temperature, the atmosphere has a random refractive-index profile.

Turbulent flow is a nonlinear process governed by the Navier-Stokes equations.

Because there are difficulties in solving the Navier-Stokes equations for fully developed turbulence, Kolmogorov developed a statistical theory. He suggested that in turbulent flow, the kinetic energy in large eddies is transferred into smaller eddies. The average size of the largest eddies, L_0, is called the outer scale. Near the ground, L_0 is on the order of the height above ground, while high above the ground, it can be just tens to hundreds of meters.[66] The average size of the smallest turbulent eddies, l_0, is called the inner scale. At very small scales, smaller than the inner scale, the energy dissipation caused by friction prevents the turbulence from sustaining itself. The inner scale l_0 can be a few millimeters near the ground to a few centimeters high above the ground.[66] The range of eddy sizes between the inner and outer scales is called the inertial subrange.

In Kolmogorov's analysis, he assumed that eddies within the inertial subrange are statistically homogeneous and isotropic within small regions of space, meaning that properties like velocity and refractive index have stationary increments. This was the reason for using the structure function rather than the more common covariance. It allowed him to use dimensional analysis to determine that the average speed of turbulent eddies v must be related to the scale size of eddies, r, via[62]

$$v \propto r^{1/3}. \tag{9.4}$$

Then, since the structure function of speed is a square of speeds, the structure function $D_v(r)$ must follow the form

$$D_v(r) = C_v^2\, r^{2/3}, \tag{9.5}$$

where C_v is the velocity structure parameter. For laminar flow, which occurs at very small scales, the physical dependencies are slightly different, so the velocity structure function follows the form

$$D_v(r) = C_v^2\, l_0^{-4/3}\, r^2. \tag{9.6}$$

For the largest scales of turbulence, the flow is highly anisotropic. If the velocity field was homogeneous and isotropic, the structure function would asymptotically approach twice the velocity variance.

This velocity framework lead to a similar analysis of potential temperature θ (potential temperature is linearly related to ordinary temperature T). The results are $\theta \propto r^{1/3}$ so that the potential temperature structure function $D_\theta(r)$ follows the same dependence as the velocity structure function, yielding[63,64]

$$D_\theta(r) = \begin{cases} C_\theta^2\, l_0^{-4/3}\, r^2, & 0 \le r \ll l_0 \\ C_\theta^2\, r^{2/3}, & l_0 \ll r \ll L_0, \end{cases} \tag{9.7}$$

where C_θ^2 is the structure parameter of θ.

A few more considerations produce a model for refractive-index statistics. Now, the refractive index at a point in space $\mathbf{r}$ can be written as

$$n(\mathbf{r}) = \mu_n(\mathbf{r}) + n_1(\mathbf{r}), \tag{9.8}$$

where $\mu_n(\mathbf{r}) \cong 1$ is the slowly varying mean value of the refractive index, and $n_1(\mathbf{r})$ is the deviation of the index from its mean value. Writing the refractive index this way creates a zero-mean random process $n_1(\mathbf{r})$, which is easier to work with for the following statistical analysis. At optical wavelengths, the refractive index of air is given approximately by

$$n(\mathbf{r}) = 1 + 77.6 \times 10^{-6}\left(1 + 7.52 \times 10^{-3}\lambda^{-2}\right)\frac{P(\mathbf{r})}{T(\mathbf{r})} \tag{9.9}$$

$$\cong 1 + 7.99 \times 10^{-5}\frac{P(\mathbf{r})}{T(\mathbf{r})} \qquad \text{for} \qquad \lambda = 0.5\,\mu\text{m}, \tag{9.10}$$

where λ is the optical wavelength in micrometers, P is the pressure in millibars, and T is the ordinary temperature in Kelvin. The variation in refractive index is given by

$$dn = 7.99 \times 10^{-5}\left(dP - \frac{-dT}{T^2}\right). \tag{9.11}$$

In this model, each eddy is considered to have relatively uniform pressure. Also, the reader should recall that potential temperature θ is linearly related to ordinary temperature T. Therefore, the refractive index variation becomes

$$dn = 7.99 \times 10^{-5}\frac{d\theta}{T^2}. \tag{9.12}$$

Because the variation in refractive index is directly proportional to the variation in potential temperature, the refractive index structure function $D_n(r)$ follows the same power law as $D_\theta(r)$ so that

$$D_n(r) = \begin{cases} C_n^2\, l_0^{-4/3}\, r^2, & 0 \le r \ll l_0 \\ C_n^2\, r^{2/3}, & l_0 \ll r \ll L_0, \end{cases} \tag{9.13}$$

where C_n^2 is known as the refractive-index structure parameter, measured in $\text{m}^{-2/3}$. It is related to the temperature structure constant by

$$C_n^2 = \left[77.6 \times 10^{-6}\left(1 + 7.52 \times 10^{-3}\lambda^{-2}\right)\frac{P}{T^2}\right]^2 C_T^2. \tag{9.14}$$

Typical values of C_n^2 are in the range 10^{-17}–10^{-13} $\text{m}^{-2/3}$, with small values at high altitudes and large values near the ground.

It is often necessary to have a spectral description of refractive-index fluctuations. The power spectral density $\Phi_n(\kappa)$ can easily be computed from Eq. (9.13) and vice versa.[15] For example, the Kolmogorov refractive-index power spectral density is computed by

$$\Phi_n^K(\kappa) = \frac{1}{4\pi^2\kappa^2}\int_0^\infty \frac{\sin(\kappa r)}{\kappa r}\frac{d}{dr}\left[r^2\frac{d}{dr}D_n(r)\right]dr \tag{9.15}$$

$$= 0.033\, C_n^2\kappa^{-11/3} \qquad \text{for} \qquad \frac{1}{L_0} \ll \kappa \ll \frac{1}{l_0}, \tag{9.16}$$

where $\boldsymbol{\kappa} = 2\pi\left(f_x\hat{\mathbf{i}} + f_y\hat{\mathbf{j}}\right)$ is angular spatial frequency in rad/m. The reader should note that Eq. (9.15) is valid only for random fields that are locally homogeneous and isotropic.

There are other models for the refractive power spectral density, like the Tatarskii, von Kármán, modified von Kármán, and Hill spectrum, which are commonly used.[15] These are each more sophisticated and include various inner-scale and outer-scale factors that improve the agreement between theory and experimental measurements. These power spectra are shown in Fig. 9.1. Two of the simplest practical models are the von Kármán PSD, given by

$$\Phi_n^{vK}(\kappa) = \frac{0.033\, C_n^2}{\left(\kappa^2 + \kappa_0^2\right)^{11/6}} \qquad \text{for} \qquad 0 \le \kappa \ll 1/l_0, \tag{9.17}$$

and the modified von Kármán PSD

$$\Phi_n^{mvK}(\kappa) = 0.033\, C_n^2\frac{\exp\left(-\kappa^2/\kappa_m^2\right)}{\left(\kappa^2 + \kappa_0^2\right)^{11/6}} \qquad \text{for} \qquad 0 \le \kappa < \infty, \tag{9.18}$$

where $\kappa_m = 5.92/l_0$ and $\kappa_0 = 2\pi/L_0$. The values of κ_m and κ_0 are chosen to match the small-scale (high-frequency) and large-scale (low-frequency) behavior predicted by the dimensional analysis. The modified von Kármán is the simplest PSD model that includes effects of both inner and outer scales.[15] Note that when $l_0 = 0$ and $L_0 = \infty$ are used, Eq. (9.18) reduces to Eq. (9.16).

When dealing with electromagnetic propagation through the atmosphere, the refractive index can be considered independent of time over short (100 μs) time scales. Because the speed of light is so fast, the time it takes light to traverse even a very large turbulent eddy is much, much shorter than the time it takes for an eddy's properties to change. Consequently, temporal properties are built into turbulence models through the Taylor frozen-turbulence hypothesis. The hypothesis is that temporal variations in meteorological quantities at a location in space are caused by advection of these quantities by the mean-speed wind flow, not by changes in the quantities themselves.[15] Consequently, turbulent eddies are treated as frozen in space and blown across the optical axis by the mean wind velocity $\mathbf{v}$. Then, with knowledge of the mean wind speed, one converts spatial statistics into temporal

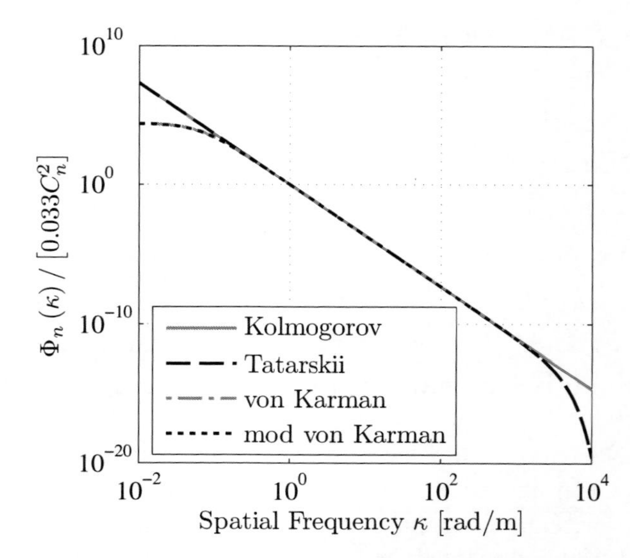

Figure 9.1 Common models for atmospheric power spectra.

statistics. For example, the temporal dependence of optical phase $\phi(x, y)$ is given by

$$\phi(x, y, t) = \phi(x - v_x t, y - v_y t, 0), \tag{9.19}$$

where v_x and v_y are the Cartesian components of the mean wind velocity, and t is time.

9.2.2 Optical propagation through turbulence

As described in Ch. 1, electromagnetic phenomena are governed by Maxwell's equations for both vacuum and atmospheric turbulence. The atmosphere may be considered a source-free, nonmagnetic, and isotropic medium. For optical-wave propagation, we seek solutions of a traveling wave with harmonic time dependence $\exp(-i2\pi\nu t)$, where $\nu = c/\lambda$ is the frequency of the light just like in Sec. 1.2.1. Then, the wave equation for the electric field may be written as[15]

$$\nabla^2 \mathbf{E}(\mathbf{r}) + k^2 n^2(\mathbf{r}) \mathbf{E}(\mathbf{r}) + 2\nabla [\mathbf{E}(\mathbf{r}) \cdot \nabla \ln n(\mathbf{r})] = \mathbf{0}, \tag{9.20}$$

where $\mathbf{E}$ is the electric field vector and k is the vacuum optical wavenumber. The last term in Eq. (9.20) refers to the change in polarization as the wave propagates. It can be neglected for $\lambda < l_0$, and consequently the wave equation simplifies to

$$\left[\nabla^2 + k^2 n^2(\mathbf{r})\right] \mathbf{E}(\mathbf{r}) = \mathbf{0}. \tag{9.21}$$

Like in Sec. 1.2.1, the magnetic induction $\mathbf{B}$ obeys this equation, too, so we can write one equation for any of the six field components:

$$\left[\nabla^2 + k^2 n^2(\mathbf{r})\right] U(\mathbf{r}) = 0. \tag{9.22}$$

This is almost identical to Eq. (1.43), except that the refractive index is explicitly position-dependent here. In solving Eq. (9.22), we recall Eq. (9.8) and assume that $|n_1(\mathbf{r})| \ll 1$. This is the assumption of weak fluctuations, which is quantified later in this chapter. With this approximation, the factor $n^2(\mathbf{r})$ in Eq. (9.22) can be approximated by

$$n^2(\mathbf{r}) \cong 1 + 2n_1(\mathbf{r}). \tag{9.23}$$

Then, the wave equation becomes

$$\left\{\nabla^2 + k^2\left[1 + 2n_1(\mathbf{r})\right]\right\} U(\mathbf{r}) = 0. \tag{9.24}$$

When the medium has a constant index of refraction, Eq. (9.22) is solved by the methods of Fourier optics from Sec. 1.3, which involve the use of Green's functions. However, when the medium is randomly inhomogeneous, as is the case with the atmosphere, perturbative methods are used with Green's functions to obtain approximate solutions. In the Rytov method, the optical field is written as

$$U(\mathbf{r}) = U_0(\mathbf{r}) \exp\left[\psi(\mathbf{r})\right], \tag{9.25}$$

where $U_0(\mathbf{r})$ is the vacuum solution ($n_1 = 0$) of Eq. (9.24), and $\psi(\mathbf{r})$ is the complex phase perturbation. The form

$$\psi(\mathbf{r}) = \psi_1(\mathbf{r}) + \psi_2(\mathbf{r}) + \ldots \tag{9.26}$$

is used to perform successive perturbations. These successive perturbations are used to compute various statistical moments of ψ which, in turn, yield statistical moments of the field. Further, it is useful to isolate amplitude and phase quantities by writing

$$\psi = \chi + i\phi, \tag{9.27}$$

where χ is the log-amplitude perturbation, and ϕ is the phase perturbation. The Rytov method can be used with a given PSD model to analytically compute moments of the field for simple source fields like Gaussian beams, spherical waves, and plane waves. The reader is referred to Clifford,[67] Ishimaru,[65] Andrews and Phillips,[15] and Sasiela[68] for greater detail about the Rytov method.

9.2.3 Optical parameters of the atmosphere

The details of the derivations are omitted here, but useful field moments that can be calculated from Rytov theory include

- the mean value of the optical field

$$\langle U(\mathbf{r})\rangle = U_0(\mathbf{r}) \langle \exp\psi(\mathbf{r})\rangle, \quad \text{and} \tag{9.28}$$

- the mutual coherence function

$$\Gamma\left(\mathbf{r},\mathbf{r}',z\right) = \left\langle U\left(\mathbf{r}\right)U^*\left(\mathbf{r}'\right)\right\rangle \tag{9.29}$$
$$= U_0\left(\mathbf{r}\right)U_0^*\left(\mathbf{r}'\right)\left\langle \exp\left[\psi\left(\mathbf{r}\right)\psi^*\left(\mathbf{r}'\right)\right]\right\rangle . \tag{9.30}$$

From the mutual coherence function, we can compute many useful properties, including

- the modulus of the complex coherence factor (hereafter called the coherence factor)[6]

$$\mu\left(\mathbf{r},\mathbf{r}',z\right) = \frac{\left|\Gamma\left(\mathbf{r},\mathbf{r}',z\right)\right|}{\left[\Gamma\left(\mathbf{r},\mathbf{r},z\right)\Gamma\left(\mathbf{r}',\mathbf{r}',z\right)\right]^{1/2}}, \tag{9.31}$$

- the wave structure function

$$D\left(\mathbf{r},\mathbf{r}',z\right) = -2\ln\mu\left(\mathbf{r},\mathbf{r}',z\right) \tag{9.32}$$
$$= D_\chi\left(\mathbf{r},\mathbf{r}',z\right) + D_\phi\left(\mathbf{r},\mathbf{r}',z\right), \tag{9.33}$$

  where D_χ and D_ϕ are the log-amplitude and phase structure functions, respectively,

- the phase power spectral density

$$\Phi_\phi\left(\kappa\right) = \frac{1}{4\pi^2\kappa^2}\int_0^\infty \frac{\sin\left(\kappa r\right)}{\kappa r}\frac{d}{dr}\left[r^2\frac{d}{dr}D_\phi\left(r\right)\right]dr, \quad \text{and} \tag{9.34}$$

- the mean MTF of the turbulent path

$$\mathcal{H}\left(f\right) = \exp\left[-\frac{1}{2}D\left(\lambda f_l f\right)\right], \tag{9.35}$$

  where f_l is the system focal length.

Each of these properties are discussed below. Then later, some of these theoretical properties are used to validate turbulent wave-optics simulations.

The structure parameter C_n^2 is a measure of the local turbulence strength. However, there are other, more useful and measurable quantities that have more intuitive meanings. Additionally, C_n^2 is a function of the propagation distance Δz, so sometimes single numbers are more handy to characterize specific optical effects. Consequently, $C_n^2\left(z\right)$ is commonly used to compute parameters like the atmospheric coherence diameter r_0 and isoplanatic angle θ_0, discussed below. In fact, the coherence diameter and isoplanatic angle are related to integrals of $C_n^2\left(z\right)$.

In the case of an isotropic and homogeneous optical field, the modulus of the coherence factor can be computed as[68]

$$\mu\left(\mathbf{r},\mathbf{r}',z\right) = \mu\left(\mathbf{r},\mathbf{r}+\Delta\mathbf{r},z\right) = \mu\left(\Delta\mathbf{r},z\right) = \mu\left(\left|\Delta\mathbf{r}\right|,z\right). \tag{9.36}$$

The exact form of the coherence factor depends on both the type of optical source and the type of refractive-index PSD being used. As a simple example, when the source is a plane wave,

$$\mu\left(|\Delta\mathbf{r}|,\Delta z\right)=\exp\left\{-4\pi^2k^2\int_0^{\Delta z}\int_0^{\infty}\Phi_n\left(\kappa,z\right)\left[1-J_0\left(\kappa\left|\Delta\mathbf{r}\right|\right)\right]\kappa\,d\kappa dz\right\},\tag{9.37}$$

and the only dependence on the propagation path is $C_n^2(z)$ within the refractive-index PSD. When the Kolmogorov spectrum is used, the coherence factor evaluates to

$$\mu^K\left(|\Delta\mathbf{r}|,\Delta z\right)=\exp\left[-1.46k^2\left|\Delta\mathbf{r}\right|^{5/3}\int_0^{\Delta z}C_n^2(z)\,dz\right].\tag{9.38}$$

The spatial coherence radius ρ_0 of an optical wave is defined as the e^{-1} point of $\mu\left(|\Delta\mathbf{r}|,\Delta z\right)$. Now, recalling Eq. (9.32) allows us to write

$$D\left(\rho_0,z\right)=2\,\mathrm{rad}^2\tag{9.39}$$

as an equivalent definition of ρ_0. With either definition, the coherence radius for a plane wave in Kolmogorov turbulence is computed as

$$\rho_0=\left[1.46k^2\int_0^{\Delta z}C_n^2(z)\,dz\right]^{-3/5}.\tag{9.40}$$

The atmospheric coherence diameter r_0 is a more commonly used parameter, and it is given by[15]

$$D\left(r_0,z\right)=6.88\,\mathrm{rad}^2\qquad\text{and}\qquad r_0=2.1\,\rho_0\tag{9.41}$$

for a plane wave. It also known as the Fried parameter because it was first introduced by D. L. Fried.[69] In fact, it was introduced in a very different way from ρ_0. Fried analyzed the resolution of an imaging telescope as the volume underneath the atmospheric MTF. When written as a function of telescope diameter, the knee in the curve was defined as r_0. For a plane-wave source, the atmospheric coherence diameter $r_{0,pw}$ is mathematically computed as[68]

$$r_{0,pw}=\left[0.423k^2\int_0^{\Delta z}C_n^2(z)\,dz\right]^{-3/5},\tag{9.42}$$

where light propagates from the source at $z=0$ to the receiver at $z=\Delta z$. For a point source (spherical wave), the atmospheric coherence diameter $r_{0,sw}$ is computed as[68]

$$r_{0,sw}=\left[0.423k^2\int_0^{\Delta z}C_n^2(z)\left(\frac{z}{\Delta z}\right)^{5/3}dz\right]^{-3/5}.\tag{9.43}$$

Values of r_0 are typically 5–10 cm for visible wavelengths and vertical viewing.

With these definitions and letting $r = |\Delta\mathbf{r}|$, the wave structure function for a plane-wave source with Kolmogorov turbulence can be written as[15]

$$D^K(r) = 6.88\left(\frac{r}{r_0}\right)^{5/3}. \tag{9.44}$$

Recall that the inner scale and outer scale are assumed to be $l_0 = 0$ and $L_0 = \infty$ in this case. Using the von Kármán PSD, we can account for a finite outer scale, resulting in a more accurate structure function given by

$$D^{vK}(r) = 6.16 r_0^{-5/3}\left[\frac{3}{5}\kappa_0^{-5/3} - \frac{(r/\kappa_0/2)^{5/6}}{\Gamma(11/6)} K_{5/6}(\kappa_0 r)\right]. \tag{9.45}$$

When both the inner and outer scales are important, we can use the modified von Kármán PSD to yield

$$D^{mvK}(r) = 3.08 r_0^{-5/3} \times\left\{\Gamma\left(-\frac{5}{6}\right)\kappa_m^{-5/3}\left[1 - {}_1F_1\left(-\frac{5}{6};1;-\frac{\kappa_m^2 r^2}{4}\right)\right] - \frac{9}{5}\kappa_0^{1/3} r^2\right\}, \tag{9.46}$$

where ${}_1F_1(a;c;z)$ is a confluent hypergeometric function of the first kind and the modified von Kármán PSD has been used. Andrews *et al.*[70] presented an algebraic approximation for the hypergeometric function that allows this structure function to be written in the simpler form

$$D^{mvK}(r) \simeq 7.75 r_0^{-5/3} l_0^{-1/3} r^2\left[\frac{1}{\left(1 + 2.03 r^2/l_0^2\right)^{1/6}} - 0.72(\kappa_0 l_0)^{1/3}\right], \tag{9.47}$$

with $< 2\%$ error. The wave structure functions for other sources and more sophisticated PSD models like the Hill model can be found in references like Andrews and Phillips.[15] The plane-wave cases are given here because they are very useful, particularly for verifying the properties of randomly generated phase screens used in wave-optics simulations.

With the various forms of the wave structure function calculated, Eq. (9.34) allows us to compute the phase PSD. Practically speaking though, there is another relationship that makes the phase PSD much easier to calculate. For a plane wave in weak turbulence, the phase PSD is

$$\Phi_\phi(\kappa) = 2\pi^2 k^2 \Delta z \Phi_n(\kappa). \tag{9.48}$$

Then, it is straightforward to show that the phase PSDs for the Kolmogorov, von Kármán, and modified von Kármán refractive-index PSD's are

$$\Phi_\phi^K(\kappa) = 0.49 r_0^{-5/3}\kappa^{-11/3}, \tag{9.49}$$

$$\Phi_{\phi}^{vK}(\kappa) = \frac{0.49 r_0^{-5/3}}{\left(\kappa^2 + \kappa_0^2\right)^{11/6}}, \tag{9.50}$$

and

$$\Phi_{\phi}^{mvK}(\kappa) = 0.49 r_0^{-5/3} \frac{\exp\left(-\kappa^2/\kappa_m^2\right)}{\left(\kappa^2 + \kappa_0^2\right)^{11/6}}, \tag{9.51}$$

respectively. Later in the chapter, these PSDs are used to generate random draws of turbulent phase screens. The method makes use of FTs, and this book's FT convention uses ordinary frequency in cycles/m, rather than angular frequency in rad/m. Accordingly, it is useful to write the PSD in terms of f, which yields

$$\Phi_{\phi}^{K}(f) = 0.023 r_0^{-5/3} f^{-11/3}, \tag{9.52}$$

as one example. The other PSDs follow similarly.

When Fried introduced r_0, he did it as a part of calculating the average MTF of images taken through the atmosphere.[69] His results can be summarized as[6]

$$\mathcal{H}(f) = \exp\left\{-3.44\left(\frac{\lambda f_l f}{r_0}\right)^{5/3}\left[1 - \alpha\left(\frac{\lambda f_l f}{D}\right)^{1/3}\right]\right\} \tag{9.53}$$

$$= \exp\left\{-3.44\left(\frac{f}{2f_0}\frac{D}{r_0}\right)^{5/3}\left[1 - \alpha\left(\frac{f}{2f_0}\right)^{1/3}\right]\right\}, \tag{9.54}$$

where again f_0 is the diffraction-limited cutoff frequency and

$$\alpha = \begin{cases} 0 & \text{for long-exposure imagery,} \\ 1 & \text{for short-exposure imagery without scintillation,} \\ \frac{1}{2} & \text{for short-exposure imagery with scintillation.} \end{cases} \tag{9.55}$$

The key distinction between short exposures and long exposures here lies in the correction of atmospheric tilt. Long-exposure images are assumed to be long enough that the image center wanders randomly many times in the image plane. Conversely, short-exposure images are assumed to be short enough that only one realization of tilt affects the image. When multiple short-exposure images are averaged, the images are first shifted to the center, thereby removing the effects of tilt. The reader should note that the atmosphere has a transfer function given by Eq. (9.54), while the imaging system has its own OTF as discussed in Sec. 5.2.2. The OTF of the composite system is the product of the two OTFs. As an example, a plot of the composite MTFs is shown in Fig. 9.2 for a circular aperture and $D/r_0 = 4$.

As discussed in Sec. 5.2.3, the average MTF can be used to determine an imaging system's Strehl ratio. Fried's work provides a way to include the effects of

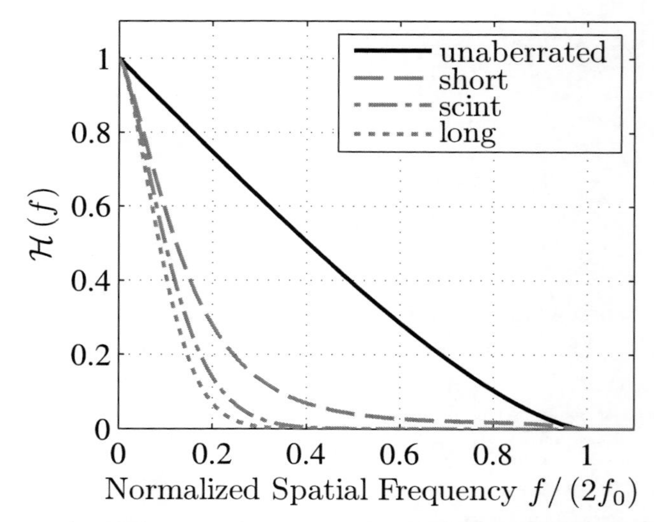

Figure 9.2 Composite MTFs for $D/r_0 = 4$. The solid black line shows the unaberrated case. The gray dashed line shows the short-exposure case with only phase fluctuations. The gray dash-dot line shows the short-exposure case when scintillation is significant. The gray dotted line shows the long-exposure case.

turbulence when calculating Strehl ratio. Making use of Eqs. (5.47) and (9.54), the Strehl ratio for a circular aperture in turbulence is given by

$$\mathcal{S} = \frac{16}{\pi} \int_0^1 f' \left(\cos^{-1} f' - f' \sqrt{1 - f'^2} \right) \times \exp \left\{ -3.44 \left(f' \frac{D}{r_0} \right)^{5/3} \left[1 - \alpha \left(f' \right)^{1/3} \right] \right\} df', \tag{9.56}$$

where $f' = f/(2f_0)$ is normalized spatial frequency. Fried numerically evaluated this integral for each value of α. Later, Andrews and Phillips developed an analytic approximation for the long-exposure case without scintillation ($\alpha = 0$) given by[15]

$$\mathcal{S} \cong \frac{1}{\left[1 + (D/r_0)^{5/3} \right]^{6/5}}. \tag{9.57}$$

Their approximation is quite accurate for all D/r_0. Sasiela evaluated this case of the integral using Mellin transforms, resulting in an expression that can be written either as a Meijer G-function or equivalently as a Fox H-function.[68] Using the first few terms of a series representation leads to the approximate polynomial expression:

$$\mathcal{S} \cong \left(\frac{r_0}{D} \right)^2 - 0.6159 \left(\frac{r_0}{D} \right)^3 + 0.0500 \left(\frac{r_0}{D} \right)^5 + 0.132 \left(\frac{r_0}{D} \right)^7, \tag{9.58}$$

which is extremely accurate for $D/r_0 > 2$.

If an optical system's characteristics (optical transfer function and point-spread function) are not shift-invariant, the system has a property called anisoplanatism. This applies to any optical system, but the system of interest here is the atmosphere. To measure the severity of angular anisoplanatism, we can examine an angular structure function of the phase $D_\phi(\theta)$ defined by

$$D_\phi(\Delta\theta) = \left\langle |\phi(\theta) - \phi(\theta + \Delta\theta)|^2 \right\rangle, \qquad (9.59)$$

where θ is an angular coordinate in the object field and $\Delta\theta$ is an angular separation between two points in the object field. The isoplanatic angle θ_0 is defined as the angle for which

$$D_\phi(\theta_0) = 1\,\text{rad}^2. \qquad (9.60)$$

By similar mathematics to those that lead to Eq. (9.43), θ_0 is given by

$$\theta_0 = \left[2.91k^2\Delta z^{5/3}\int_0^{\Delta z} C_n^2(z)\left(1-\frac{z}{\Delta z}\right)^{5/3} dz\right]^{-3/5}. \qquad (9.61)$$

This may be considered the largest field angle over which the optical path length through the turbulence does not differ significantly from the on-axis optical path length through the turbulence. Values of θ_0 are typically 5–10 μrad for visible wavelengths and vertical viewing.

Log-amplitude (or equivalently, irradiance) statistics are also important to describe the strength of scintillations. The log-amplitude variance, defined as

$$\sigma_\chi^2(\mathbf{r}) = \left\langle \chi^2(\mathbf{r}) \right\rangle - \left\langle \chi(\mathbf{r}) \right\rangle^2, \qquad (9.62)$$

is a common measure of scintillation. For plane-wave and diverging spherical-wave (point) sources, the log-amplitude variances $\sigma_{\chi,pw}^2$ and $\sigma_{\chi,sw}^2$ evaluate to[68]

$$\sigma_{\chi,pw}^2 = 0.563k^{7/6}\Delta z^{5/6}\int_0^{\Delta z} C_n^2(z)\left(1-\frac{z}{\Delta z}\right)^{5/6} dz \qquad (9.63)$$

and

$$\sigma_{\chi,sw}^2 = 0.563k^{7/6}\int_0^{\Delta z} C_n^2(z)\, z^{5/6}\left(1-\frac{z}{\Delta z}\right)^{5/6} dz, \qquad (9.64)$$

respectively. Weak fluctuations are associated with $\sigma_\chi^2 < 0.25$, and strong fluctuations with $\sigma_\chi^2 \gg 0.25$. Note that the Rytov method presented here is valid only for weak fluctuations.

9.2.4 Layered atmosphere model

Deriving analytic results for atmospheric turbulence effects on optical propagation is possible when we assume a simple statistical model. However, when one wants to consider more complex scenarios like using adaptive-optics systems, usually the statistics of the corrected optical fields cannot be computed in closed form. For mathematical simplification, a common technique is to treat turbulence as a finite number of discrete layers. This approach is common for analytic calculations, computer simulations, and emulating turbulence in the laboratory.[15,60,61] A layered model is useful if its refractive index spectrum and scintillation properties match that of the corresponding extended medium.[23,71]

Each layer is a unit-amplitude thin phase screen which represents a turbulent volume of a much greater thickness. A phase screen is considered thin if its thickness is much less than the propagation distance following the screen.[15] A phase screen is one realization of an atmospheric phase perturbation, and it is used with Eq. (9.2) to compute a realization of the refraction operator $\mathcal{T}\left[z_i, z_{i+1}\right]$. This is how atmospheric phase screens are incorporated into the split-step beam propagation method to simulate atmospheric propagation. A discussion of layered turbulence theory and phase screen generation follows.

9.2.5 Theory

To theoretically represent the atmosphere as phase screens, we simply write the turbulence profile in terms of the effective structure parameter $C_{n_i}^2$, the location along the propagation path z_i, and the thickness Δz_i of the slab of extended turbulence represented by the ith phase screen. The values of $C_{n_i}^2$ are chosen so that several low-order moments of the continuous model match the layered model:[23,71]

$$\int_0^{\Delta z} C_n^2\left(z'\right)\left(z'\right)^m dz' = \sum_{i=1}^{n} C_{n_i}^2 z_i^m \Delta z_i, \tag{9.65}$$

where n is the number of phase screens being used, and $0 \leq m \leq 7$. This way, r_0, θ_0, σ_χ^2, etc. of the layered model match the parameters of the bulk turbulence being modeled. The atmospheric parameters for the layered turbulence model are computed using the discrete-sum versions of Eqs. (9.42), (9.43), (9.63), and (9.64) given by

$$r_{0,pw} = \left(0.423k^2 \sum_i C_{n_i}^2 \Delta z_i\right)^{-3/5} \tag{9.66}$$

$$r_{0,sw} = \left[0.423k^2 \sum_{i=1}^{n} C_{n_i}^2 \left(\frac{z_i}{\Delta z}\right)^{5/3} \Delta z_i\right]^{-3/5} \tag{9.67}$$

$$\sigma_{\chi,pw}^2 = 0.563k^{7/6}\Delta z^{5/6} \sum_{i=1}^{n} C_{n_i}^2 \left(1 - \frac{z_i}{\Delta z}\right)^{5/6} \Delta z_i \tag{9.68}$$

$$\sigma_{\chi,sw}^2 = 0.563 k^{7/6} \Delta z^{5/6} \sum_{i=1}^{n} C_{n_i}^2 \left(\frac{z_i}{\Delta z}\right)^{5/6} \left(1 - \frac{z_i}{\Delta z}\right)^{5/6} \Delta z_i. \tag{9.69}$$

By grouping terms in Eq. (9.66), the ith layer can be given an effective coherence diameter r_{0_i} given by[71]

$$r_{0_i} = \left[0.423\, k^2\, C_{n_i}^2 \Delta z_i\right]^{-3/5}. \tag{9.70}$$

Note that this is the plane-wave r_0, so it is valid only when the layer is very thin. The r_0 values for turbulence layers are commonly used for characterizing their strength. With this definition, Eq. (9.70) can be substituted into Eqs. (9.66)–(9.69) to write the desired optical field properties in terms of the phase-screen r_0 values. This substitution yields

$$r_{0,pw} = \left(\sum_{i=1}^{n} r_{0_i}^{-5/3}\right)^{-3/5} \tag{9.71}$$

$$r_{0,sw} = \left[\sum_{i=1}^{n} r_{0_i}^{-5/3} \left(\frac{z_i}{\Delta z}\right)^{5/3}\right]^{-3/5} \tag{9.72}$$

$$\sigma_{\chi,pw}^2 = 1.33\, k^{-5/6} \Delta z^{5/6} \sum_{i=1}^{n} r_{0_i}^{-5/3} \left(1 - \frac{z_i}{\Delta z}\right)^{5/6} \tag{9.73}$$

$$\sigma_{\chi,sw}^2 = 1.33\, k^{-5/6} \Delta z^{5/6} \sum_{i=1}^{n} r_{0_i}^{-5/3} \left(\frac{z_i}{\Delta z}\right)^{5/6} \left(1 - \frac{z_i}{\Delta z}\right)^{5/6}. \tag{9.74}$$

Given a set of desired atmospheric conditions, $r_{0,sw}$ and $\sigma_{\chi,sw}^2$ for example, these equations could be used to determine the required phase screen properties and locations along the path. These equations could be written in matrix-vector notation. Using a typical number of phase screens, like 5–10, there are 10–20 unknown parameters (r_0 and z_i for each screen), and so the system of two equations is far underdetermined. This is easy to improve by simply fixing phase screen locations. For example, we could maintain consistency with the uniform spacing of the partial-propagation planes, as discussed in Ch. 8. Then, choosing to place a phase screen in each partial-propagation plane, we can recall from Sec. 8.3 that $\alpha_i = z_i/\Delta z$, which simplifies the equations further. As an example, the system of equations for five screens would look like

$$\begin{pmatrix} r_{0,sw}^{-5/3} \\ \frac{\sigma_{\chi,sw}^2}{1.33}\left(\frac{k}{\Delta z}\right)^{5/6} \end{pmatrix} = \begin{pmatrix} 0 & 0.0992 & 0.315 & 0.619 & 1 \\ 0 & 0.248 & 0.315 & 0.248 & 0 \end{pmatrix} \begin{pmatrix} r_{01}^{-5/3} \\ r_{02}^{-5/3} \\ r_{03}^{-5/3} \\ r_{04}^{-5/3} \\ r_{05}^{-5/3} \end{pmatrix}. \tag{9.75}$$

The entries in the first row of the matrix are $\alpha_i^{5/3}$, and the entries in the second row of the matrix are $\alpha_i^{5/6}(1-\alpha_i)^{5/6}$.

In this approach, the left side is determined by the scenario we want to simulate. Given λ, Δz, and a model of $C_n^2(z)$, we compute the desired atmospheric parameters for the simulation. Then, we solve an appropriate system of equations, like Eq. (9.75), to compute the phase screen r_0 values. The difficulty with this approach is the $-5/3$ power in the r_0 vector. Negative entries in the solved r_0 vector are unphysical, so the solutions must be constrained to positive values. The example in Sec. 9.5 shows use of constrained optimization to compute r_0 values for a simulation with several phase screens.

9.3 Monte-Carlo Phase Screens

The refractive index variation of the atmosphere is a random process, and so is the optical path length through it. Consequently, turbulence models give statistical averages, like the structure function and power spectrum of refractive index variations. The problem of creating atmospheric phase screens is one of generating individual realizations of a random process. That is, phase screens are created by transforming computer-generated random numbers into two-dimensional arrays of phase values on a grid of sample points that have the same statistics as turbulence-induced phase variations. The literature is rife with clever methods to generate atmospheric phase screens with good computational efficiency,[72–75] high accuracy,[56,71,76–82] and flexibility.[83–85]

Usually, the phase is written as a weighted sum of basis functions. The common basis sets used for this purpose have been Zernike polynomials and Fourier series. Both basis sets have benefits and drawbacks. The most common method for phase-screen generation is based on the FT, first introduced by McGlamery.[86]

Assuming that turbulence-induced phase $\phi(x,y)$ is a Fourier-transformable function, we can write it in a Fourier-integral representation as

$$\phi(x,y) = \int_{-\infty}^{\infty}\int_{-\infty}^{\infty} \Psi(f_x,f_y)\, e^{i2\pi(f_x x + f_y y)}\, df_x\, df_y, \tag{9.76}$$

where $\Psi(f_x,f_y)$ is the spatial-frequency-domain representation of the phase. Of course, $\phi(x,y)$ is actually a realization of a random process with a power spectral density given by $\Phi_\phi(f)$ [or equivalently, $\Phi_\phi(\kappa)$] as discussed in Sec. 9.2.3. Treating the phase as a two-dimensional signal, the total power P_{tot} in the phase can be written two ways using the definition of power spectral density and Parseval's theorem so that

$$P_{tot} = \int_{-\infty}^{\infty}\int_{-\infty}^{\infty} |\phi(x,y)|^2\, dx\, dy = \int_{-\infty}^{\infty}\int_{-\infty}^{\infty} \Phi_\phi(f_x,f_y)\, df_x\, df_y. \tag{9.77}$$

To generate phase screens on a finite grid, we write the optical phase $\phi(x, y)$ as a Fourier series so that[80]

$$\phi(x, y) = \sum_{n=-\infty}^{\infty} \sum_{m=-\infty}^{\infty} c_{n,m} \exp\left[i2\pi\left(f_{x_n} x + f_{y_m} y\right)\right], \tag{9.78}$$

where f_{x_n} and f_{y_m} are the discrete x- and y-directed spatial frequencies, and the $c_{n,m}$ are the Fourier-series coefficients. Because the phase variation through the atmosphere is due to many independent random inhomogeneities along the optical path, we use the central-limit theorem to determine that the $c_{n,m}$ have a Gaussian distribution. Also note that, in general, the Fourier coefficients $c_{n,m}$ are complex. The real and imaginary parts each have zero mean and equal variances, and their cross-covariances are zero. Consequently, they obey circular complex Gaussian statistics with zero mean and variance given by[32,80]

$$\left\langle \left|c_{n,m}\right|^2 \right\rangle = \Phi_\phi\left(f_{x_n}, f_{y_m}\right) \Delta f_{x_n} \Delta f_{y_m}. \tag{9.79}$$

If the FFT is to be used for computational efficiency, the frequency samples must be linearly spaced on a Cartesian grid. Then, if the x and y grid sizes are L_x and L_y, respectively, the frequency spacings are $\Delta f_{x_n} = 1/L_x$ and $\Delta f_{y_m} = 1/L_y$ so that

$$\left\langle \left|c_{n,m}\right|^2 \right\rangle = \frac{1}{L_x L_y} \Phi_\phi\left(f_{x_n}, f_{y_m}\right). \tag{9.80}$$

Now, the task is to produce realizations of the Fourier coefficients. Typical random-number software, like MATLAB's `randn` function, generates Gaussian random numbers with zero mean and unit variance. This just requires a simple transformation. If x is a Gaussian random variable with mean μ and variance σ^2, then the variable $z = (x - \mu)/\sigma$ is a Gaussian random variable with zero mean and unit variance. With this in mind, we simply generate Gaussian random numbers via standard mathematical software with zero mean and unit variance. Then, multiplication by the square root of the variance given in Eq. (9.79) produces the random draws of the FS coefficients in Eq. (9.78).

Listing 9.2 gives MATLAB code for generating phase screens using the FT method. Lines 6–16 set up the square root of Eq. (9.51). As part of the process, line 16 sets the zero-frequency component of the phase to zero. Then, line 18 generates a random draw of the FS coefficients. Finally, line 20 synthesizes the phase screen from random draws using an FT. Note that the real and imaginary parts of the IFT produce two uncorrelated phase screens. Line 20 uses the screen from the real part and discards the imaginary part.

Unfortunately, the FFT method shown in Listing 9.2 does not produce accurate phase screens. To begin understanding this, the reader should note that the phase PSDs shown in Fig. 9.1 given in Eq. (9.51) have much of the power in the low spatial frequencies. In fact, it has been well documented that we often cannot sample the spatial-frequency grid low enough to accurately represent low-order

Listing 9.2 MATLAB code for generating phase screens that are consistent with atmospheric turbulence from random draws. This code uses the FT method.

```
1  function phz = ft_phase_screen(r0, N, delta, L0, l0)
2  % function phz ...
3  %     = ft_phase_screen(r0, N, delta, L0, l0)
4
5      % setup the PSD
6      del_f = 1/(N*delta);    % frequency grid spacing [1/m]
7      fx = (-N/2 : N/2-1) * del_f;
8      % frequency grid [1/m]
9      [fx fy] = meshgrid(fx);
10     [th f] = cart2pol(fx, fy);  % polar grid
11     fm = 5.92/l0/(2*pi); % inner scale frequency [1/m]
12     f0 = 1/L0;             % outer scale frequency [1/m]
13     % modified von Karman atmospheric phase PSD
14     PSD_phi = 0.023*r0^(-5/3) * exp(-(f/fm).^2) ...
15         ./ (f.^2 + f0^2).^(11/6);
16     PSD_phi(N/2+1,N/2+1) = 0;
17     % random draws of Fourier coefficients
18     cn = (randn(N) + i*randn(N)) .* sqrt(PSD_phi)*del_f;
19     % synthesize the phase screen
20     phz = real(ift2(cn, 1));
```

modes like tilt. This difference is evident in Fig. 9.3 when we generate and verify phase screens for an example simulation through turbulence. For this figure, 40 turbulent phase screens were generated using the FT method implemented by the `ft_phase_screen` function in Listing 9.2. Then, the structure function of each screen was computed using the `str_fcn2_ft` function in Listing 3.7, and the results were averaged. A slice of the average structure function is shown by the dotted line. Clearly, the screens' statistics do not match up well with the theoretical structure function shown by the solid gray line. The poorest agreement is at large separations, which correspond to low spatial frequencies.

Several approaches have been suggested to compensate for this shortcoming. For example, Cochran,[76] Roddier,[87] and Jakobssen[79] use random draws of Zernike polynomials (or linear combinations thereof) using the Zernike-mode statistics reported by Noll.[22] In contrast, Welsh[80] and Eckert and Goda[82] use FS methods with non-uniform sampling in the spatial-frequency domain to include very low spatial frequencies. Still others use a combination of these two approaches, called "subharmonics". This approach, used by Herman and Strugala,[77] Lane *et al.*,[78] Johansson and Gavel,[88] and Sedmak,[81] augments FT screens with a low-frequency Fourier series.

Here, we implement the subharmonic method described by Lane *et al.*[78] Frehlich

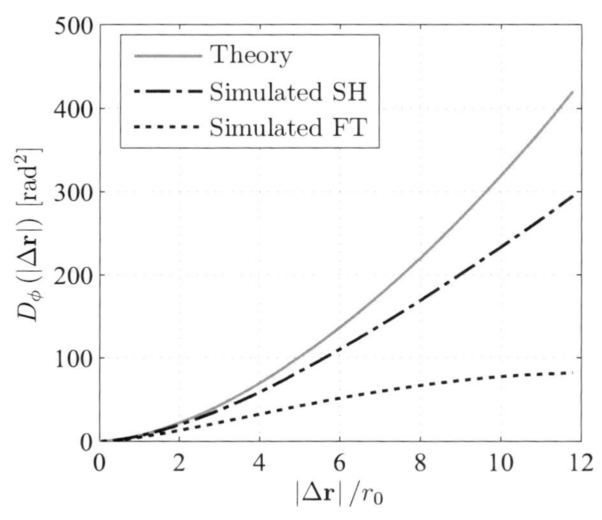

Figure 9.3 Comparison of the average structure function computed from FT and subharmonic screens against theory.

showed that turbulent simulations using these screens produce accurate results.[56] Listing 9.3 gives MATLAB code for generating phase screens using this method. In Line 8, this method begins by generating a phase screen using the FT method already discussed. Then, a low-frequency screen is generated in lines 12–37. This screen $\phi_{LF}(x, y)$ is a sum of N_p different screens, as given by

$$\phi_{LF}(x, y) = \sum_{p=1}^{N_p} \sum_{n=-1}^{1} \sum_{m=-1}^{1} c_{n,m} \exp\left[i2\pi\left(f_{x_n} x + f_{y_m} y\right)\right], \tag{9.81}$$

where the sums over n and m are over discrete frequencies and each value of the index p corresponds to a different grid. The square root of the PSD is setup in lines 14–26, the random draws of Fourier coefficients are generated in lines 28–29, and the sum over the indices n and m is carried out in lines 32–35. Then, the sum over the N_p different grids is carried out in line 36. In this particular implementation, only a 3×3 grid of frequencies is used for each value of p, and $N_p = 3$ different grids are used. The frequency grid spacing for each value of p is $\Delta f_p = 1/(3^p L)$. In this way, the frequency grids have a spacing that is a subharmonic of the FT screen's grid spacing.

Listing 9.4 gives an example of generating random phase screens using the MATLAB function `ft_sh_phase_screen` from Listing 9.3. In the listing, the screen size is 2 m, the coherence diameter is $r_0 = 10$ cm, the inner scale is $l_0 = 1$ cm, and the outer scale is $L_0 = 100$ m. An atmospheric phase-screen realization generated by Listing 9.4 is shown in Fig. 9.4.

Figure 9.3 shows verification that subharmonic screens do produce more-accurate phase screen statistics. Several authors have investigated the subharmonic

Listing 9.3 MATLAB code for generating phase screens that are consistent with atmospheric turbulence from random draws. This code uses the FT method augmented with subharmonics.

```
1  function [phz_lo phz_hi] ...
2      = ft_sh_phase_screen(r0, N, delta, L0, l0)
3  % function [phz_lo phz_hi] ...
4  %     = ft_sh_phase_screen(r0, N, delta, L0, l0)
5
6      D = N*delta;
7      % high-frequency screen from FFT method
8      phz_hi = ft_phase_screen(r0, N, delta, L0, l0);
9      % spatial grid [m]
10     [x y] = meshgrid((-N/2 : N/2-1) * delta);
11     % initialize low-freq screen
12     phz_lo = zeros(size(phz_hi));
13     % loop over frequency grids with spacing 1/(3^p*L)
14     for p = 1:3
15         % setup the PSD
16         del_f = 1 / (3^p*D); %frequency grid spacing [1/m]
17         fx = (-1 : 1) * del_f;
18         % frequency grid [1/m]
19         [fx fy] = meshgrid(fx);
20         [th f] = cart2pol(fx, fy); % polar grid
21         fm = 5.92/l0/(2*pi); % inner scale frequency [1/m]
22         f0 = 1/L0;           % outer scale frequency [1/m]
23         % modified von Karman atmospheric phase PSD
24         PSD_phi = 0.023*r0^(-5/3) * exp(-(f/fm).^2) ...
25             ./ (f.^2 + f0^2).^(11/6);
26         PSD_phi(2,2) = 0;
27         % random draws of Fourier coefficients
28         cn = (randn(3) + i*randn(3)) ...
29             .* sqrt(PSD_phi)*del_f;
30         SH = zeros(N);
31         % loop over frequencies on this grid
32         for ii = 1:9
33             SH = SH + cn(ii) ...
34                 * exp(i*2*pi*(fx(ii)*x+fy(ii)*y));
35         end
36         phz_lo = phz_lo + SH;   % accumulate subharmonics
37     end
38     phz_lo = real(phz_lo) - mean(real(phz_lo(:)));
```

Listing 9.4 Example usage of `ft_sh_phase_screen` function

```
1  % example_ft_sh_phase_screen.m
2
3  D = 2;      % length of one side of square phase screen [m]
4  r0 = 0.1; % coherence diameter [m]
5  N = 256;    % number of grid points per side
6  L0 = 100; % outer scale [m]
7  l0 = 0.01;% inner scale [m]
8
9  delta = D/N;      % grid spacing [m]
10 % spatial grid
11 x = (-N/2 : N/2-1) * delta;
12 y = x;
13 % generate a random draw of an atmospheric phase screen
14 [phz_lo phz_hi] ...
15     = ft_sh_phase_screen(r0, N, delta, L0, l0);
16 phz = phz_lo + phz_hi;
```

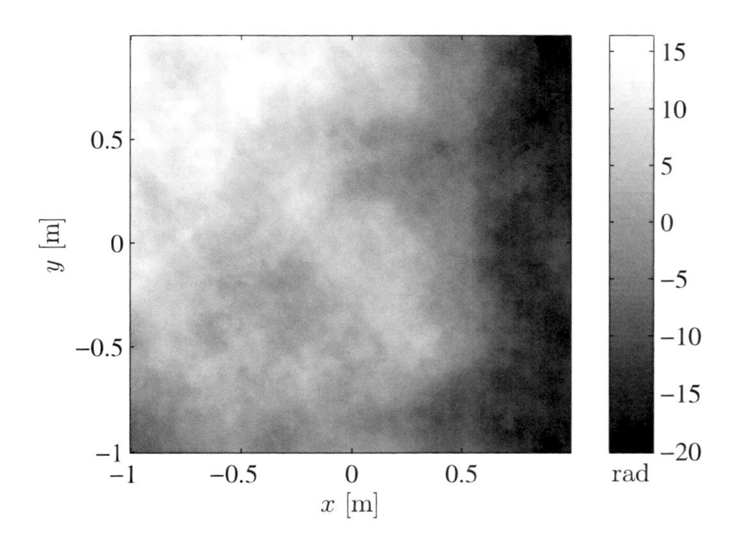

Figure 9.4 Typical atmospheric phase screen created using the subharmonic method.

method's ability to do this. Among the first to do this were Herman and Strugala.[77] While they used a slightly different version of the subharmonic method, they did show that the concept produces phase screens that result in a structure function with a good match to theory. Further, they compared the average Strehl ratio from their subharmonic screens, and it matched theory closely. Later, Lane *et al.* developed the particular subharmonic method used here and demonstrated that their screens also matched the theoretical structure function closely.[78] Shortly thereafter, Johansson and Gavel compared the approaches of Herman and Strugala and Lane *et al.*, and demonstrated their own subharmonic technique whose screens produce a structure function that matches theory very closely.[88] While investigating accuracy of non-square subharmonic phase screens, Sedmak showed good agreement with phase structure function and aperture-averaged phase variance.[81] Finally, Frehlich studied the accuracy of full wave-optics simulations making use of subharmonic screens.[56] His study showed that for beam waves, the mean irradiance is fairly accurate for both FT screens and subharmonic screens, but the subharmonic screens are far more accurate in producing the correct irradiance variance. For plane waves, both methods produced accurate irradiance variances, but only the subharmonic method produced an accurate mutual coherence function.

9.4 Sampling Constraints

As light propagates through turbulence, it spreads due to two effects: tilt and higher-order aberrations. High-order aberrations cause the beam to expand beyond the spreading due to diffraction alone. Tilt causes the beam to wander off the optical axis in a random way. Over time ($\gtrsim$ 1 msec), this random wandering causes optical energy to land all over the observation plane. Beam spreading due to high-order aberrations can be seen in a short-exposure image, whereas beam spreading due to tilt can only be seen in a long-exposure image. A full discussion of beam spreading is beyond the scope of this book, but a simple model for sampling analysis is presented below.

This turbulence-induced beam spreading makes sampling requirements even more restrictive than the vacuum constraints from Sec. 8.4. Several approaches for conducting properly sampled turbulence simulations have been discussed. For example, in vacuum propagation Johnston and Lane filter the free-space transfer function and set their grid size based the bandwidth of the filter.[41] Then, they set the sample interval based on avoiding aliasing of the quadratic phase factor just like in Sec. 7.3.2. For atmospheric simulations, they choose the grid spacing based on the phase structure function. In doing so, they compute the grid spacing δ_ϕ at which phase differences less than π in adjacent grid points occur more than 99.7% of the time. They also give consideration to sampling scintillation. The scale size of scintillation is given approximately by the Fresnel length $(\lambda\Delta z)^{1/2}$, so they set δ_i to be the smallest of δ_ϕ, $(\lambda\Delta z)^{1/2}/2$, and the grid spacing that just barely avoids aliasing of the free-space point spread function. In this way, they adequately sample free-space propagation and turbulent phase and amplitude variations. Martin and Flatté studied sampling constraints, mainly based on the PSD of the turbulence-induced irradiance fluctuations.[43] Finally, Coles *et al.* conducted a quantitative error analysis for plane waves and point sources.[32] In particular, they studied the error in observation-plane irradiance due to finite grid spacing, finite number of samples, and finite number of screens. They used only FT phase screens, so part of the error they encountered was due to the screens themselves.

Mansell, Praus, and Coy take a different approach, but one that integrates well with the frameworks presented in Chs. 7–8.[35,42,54] They modify the sampling inequalities to account for turbulence-induced beam spreading. The two sampling constraints that originate from propagation geometry are affected by turbulence. The other constraint that originates from the numerical algorithm is not affected by turbulence.

Previously, constraints 1 and 2 were stated for vacuum propagation as

$$1. \quad \delta_n \leq \frac{\lambda \Delta z - D_2 \delta_1}{D_1} \tag{9.82}$$

$$2. \quad N \geq \frac{D_1}{2\delta_1} + \frac{D_2}{2\delta_n} + \frac{\lambda \Delta z}{2\delta_1 \delta_n}. \tag{9.83}$$

Constraint 1 ensures that the source-plane grid is sampled finely enough so that all of the rays that land within the observation-plane region of interest are present in the source. In the geometric-optics approximation, turbulence causes the source's rays to refract randomly as shown in Fig. 9.5. This blurs the size of D_1 as viewed in the observation plane and the size of D_2 as viewed in the source plane. We need a model for this blurring that depends on the turbulence to adjust these two constraints.

The approach of Coy is to model the turbulence-induced beam spreading as if it were caused by a diffraction grating with period equal to r_0. This allows us to define new limiting aperture sizes D_1' and D_2' via

$$D_1' = D_1 + c\frac{\lambda \Delta z}{r_{0,rev}} \tag{9.84}$$

$$D_2' = D_2 + c\frac{\lambda \Delta z}{r_0}, \tag{9.85}$$

where $r_{0,rev}$ is the coherence diameter computed for light propagating in reverse, i.e., from the observation plane to the source plane, and c is an adjustable parameter indicating the sensitivity of the model to the turbulence. Typical values of c range from 2 to 8. Choosing $c = 2$ typically captures $\sim$97% of the light, and choosing $c = 4$ typically captures $\sim$99% of the light. Now, for simulating propagation through turbulence, the required sampling analysis utilizes the following inequalities:

$$1. \quad \delta_n \leq \frac{\lambda \Delta z - D_2' \delta_1}{D_1'} \tag{9.86}$$

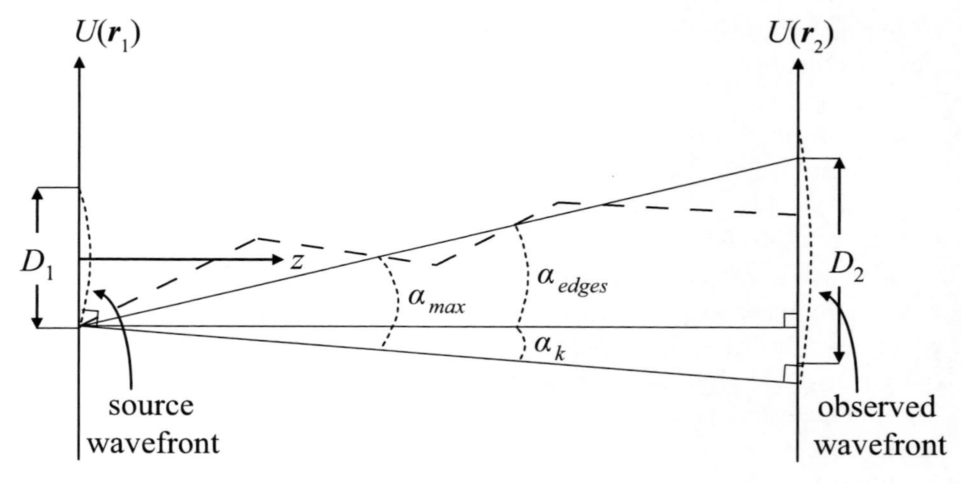

Figure 9.5 Propagation geometry in which turbulence refracts rays as the light propagates indicated by the dashed ray. This geometry leads to Constraint 1.

2. $$N \geq \frac{D_1'}{2\delta_1} + \frac{D_2'}{2\delta_n} + \frac{\lambda \Delta z}{2\delta_1 \delta_n} \tag{9.87}$$

3. $$\left(1 + \frac{\Delta z}{R}\right)\delta_1 - \frac{\lambda \Delta z}{D_1} \leq \delta_2 \leq \left(1 + \frac{\Delta z}{R}\right)\delta_1 + \frac{\lambda \Delta z}{D_1}. \tag{9.88}$$

Then, once N, δ_1, and δ_n are chosen, the partial propagation distances and number of partial propagations are chosen from

$$\Delta z_{max} = \frac{\min(\delta_1, \delta_n)^2 N}{\lambda} \tag{9.89}$$

$$n_{min} = \mathrm{ceil}\left(\frac{\Delta z}{\Delta z_{max}}\right) + 1, \tag{9.90}$$

as before.

9.5 Executing a Properly Sampled Simulation

As in Chs. 7–8, the most effective way to illustrate application of the above sampling constraints is by example. The remainder of this section illustrates the steps involved in setting up a simulation of optical-wave propagation through atmospheric turbulence.

9.5.1 Determine propagation geometry and turbulence conditions

The example simulation in this subsection is for a point source propagating a distance $\Delta z = 50$ km through a turbulent path with $C_n^2 = 1 \times 10^{-16}$ m$^{-2/3}$ along the

Listing 9.5 MATLAB code for setting up source and receiver geometry and turbulence-related quantities.

```
1  % example_pt_source_atmos_setup.m
2
3  % determine geometry
4  D2 = 0.5;     % diameter of the observation aperture [m]
5  wvl = 1e-6;  % optical wavelength [m]
6  k = 2*pi / wvl; % optical wavenumber [rad/m]
7  Dz = 50e3;       % propagation distance [m]
8
9  % use sinc to model pt source
10 DROI = 4 * D2;  % diam of obs-plane region of interest [m]
11 D1 = wvl*Dz / DROI;     % width of central lobe [m]
12 R = Dz; % wavefront radius of curvature [m]
13
14 % atmospheric properties
15 Cn2 = 1e-16;     % structure parameter [m^-2/3], constant
16 % SW and PW coherence diameters [m]
17 r0sw = (0.423 * k^2 * Cn2 * 3/8 * Dz)^(-3/5);
18 r0pw = (0.423 * k^2 * Cn2 * Dz)^(-3/5);
19 p = linspace(0, Dz, 1e3);
20 % log-amplitude variance
21 rytov = 0.563 * k^(7/6) * sum(Cn2 * (1-p/Dz).^(5/6) ...
22     .* p.^(5/6) * (p(2)-p(1)));
23
24 % screen properties
25 nscr = 11; % number of screens
26 A = zeros(2, nscr); % matrix
27 alpha = (0:nscr-1) / (nscr-1);
28 A(1,:) = alpha.^(5/3);
29 A(2,:) = (1 - alpha).^(5/6) .* alpha.^(5/6);
30 b = [r0sw.^(-5/3); rytov/1.33*(k/Dz)^(5/6)];
31 % initial guess
32 x0 = (nscr/3*r0sw * ones(nscr, 1)).^(-5/3);
33 % objective function
34 fun = @(X) sum((A*X(:) - b).^2);
35 % constraints
36 x1 = zeros(nscr, 1);
37 rmax = 0.1; % maximum Rytov number per partial prop
38 x2 = rmax/1.33*(k/Dz)^(5/6) ./ A(2,:);
39 x2(A(2,:)==0) = 50^(-5/3)
40 [X,fval,exitflag,output] ...
41     = fmincon(fun,x0,[],[],[],[],x1,x2)
42 % check screen r0s
43 r0scrn = X.^(-3/5)
44 r0scrn(isinf(r0scrn)) = 1e6;
45 % check resulting r0sw & rytov
46 bp = A*X(:); [bp(1)^(-3/5) bp(2)*1.33*(Dz/k)^(5/6)]
47 [r0sw rytov]
```

entire path. For simplicity, we assume that the Kolmogorov refractive-index PSD is adequate for our purposes. The telescope observing the light is $D_2 = 0.5$ m in diameter. With this information, we can compute the atmospheric parameters of interest. This, of course, depends on what we want to do with the light after propagation. Perhaps we may want to do imaging, wavefront sensing, adaptive optics, and more. In this particular example, we are simply interested in verifying that the simulation is operating correctly. To verify, we propagate the source through many realizations of turbulence, compute the coherence factor, and plot it against the theoretical expectation. We also need to determine the locations of the phase screens and their coherence diameters.

Listing 9.5 gives the MATLAB code for setting up the turbulence model. This starts with setting aperture sizes, optical wavelength, propagation distance, etc. Lines 10–11 compute D_1 from the width of the model point source's central lobe. This begins with setting the diameter of the region of interest (the variable `DROI`) that is uniformly illuminated in the observation plane by the source. Lines 17–22 continue with computing the key atmospheric parameters, $r_{0,sw} = 12.7$ cm and $\sigma^2_{\chi,sw} = 0.436$, from Eqs. (9.43) and (9.64), respectively.

Lines 25–41 compute the phase screen r_0 values according to the approach in Sec. 9.2.5. In this process, lines 26–29 set up the matrix, which is similar to the matrix in Eq. (9.75). Line 30 sets up the vector, which is the left side in Eq. (9.75) (the variable `b`). With the known matrix and vector determined, the screen r_0 values must be computed through a constrained search through possible values of screen r_0's. Actually, the parameters are the $-5/3$ power of the screen r_0's in the variable `X` according to Eq. (9.75). Their values are computed through constructing an objective function that can be minimized when suitable r_0 values are found within a valid range. This objective function in line 34 is the difference between the desired atmospheric parameters (the variable `b`) and those arising from a given choice of r_0 values (`A*X(:)`). The valid range of the `X` values is determined in lines 36–39. The lower bound of `X` is zero, corresponding to infinite screen r_0's. The upper bound is set by requiring that each screen's contribution to the overall Rytov number is less than 0.1 (see line 37). This is related to a guideline suggested by Martin and Flatté.[43] Finally, lines 40–41 perform the search to minimize the objective function, and lines 46–47 compute the atmospheric parameters based on the solved screen r_0's and print them to the command line.

9.5.2 Analyze the sampling constraints

Once the geometry and turbulence conditions are set up, we can analyze the sampling constraints to determine the grid spacings and number of grid points. In Listing 9.6, we evaluate Eqs. (9.86)–(9.88) and perform a sampling analysis using essentially the same method as in Sec. 8.4. Lines 2–16 evaluate the bounds of constraints 1–3. This is used to produce the contour plot shown in Fig. 9.6, although the plotting code is not shown. The figure shows the lower bound on N in constraint 2

Listing 9.6 MATLAB code for analyzing sampling constraints given the geometry and turbulence conditions.

```
1  % analysis_pt_source_atmos_samp.m
2  c = 2;
3  D1p = D1 + c*wvl*Dz/r0sw;
4  D2p = D2 + c*wvl*Dz/r0sw;
5
6  delta1 = linspace(0, 1.1*wvl*Dz/D2p, 100);
7  deltan = linspace(0, 1.1*wvl*Dz/D1p, 100);
8  % constraint 1
9  deltan_max = -D2p/D1p*delta1 + wvl*Dz/D1p;
10 % constraint 3
11 d2min3 = (1+Dz/R)*delta1 - wvl*Dz/D1p;
12 d2max3 = (1+Dz/R)*delta1 + wvl*Dz/D1p;
13 [delta1 deltan] = meshgrid(delta1, deltan);
14 % constraint 2
15 N2 = (wvl * Dz + D1p*deltan + D2p*delta1) ...
16      ./ (2 * delta1 .* deltan);
17 % constraint 4
18 d1 = 10e-3;
19 d2 = 10e-3;
20 N = 512;
21 d1*d2 * N / wvl
22 zmax = min([d1 d2])^2 * N / wvl
23 nmin = ceil(Dz / zmax) + 1
```

with the upper bound from constraints 1 and 3 overlayed. This allows us to choose the grid spacings δ_1 and δ_n in the source and observation planes, respectively, and the minimum required number of grid points, N. Then, given our choices for δ_1, δ_n, and N, we can compute the maximum allowed propagation distance Δz_{max} using Eq. (9.89) and then corresponding number of partial propagations, $n-1$, using Eq. (9.90).

The results of the analysis are given in lines 18–23, which assumes that we have already made the plots and viewed them. The chosen grid spacings are $\delta_1 = 1$ cm, and $\delta_n = 1$ cm. This gives five samples across the central peak of the model point source and 50 samples across the observing telescope aperture. This is marked on Fig. 9.6 with a white $\times$. We can see that these spacings easily satisfy constraints 1 and 3. Also, the required number of grid points is more than 2^8, so we pick $2^9 = 512$ grid points. Finally, the minimum number of planes is two, so we could use just one propagation. However, we use ten propagations (11 planes) to represent the atmosphere properly.

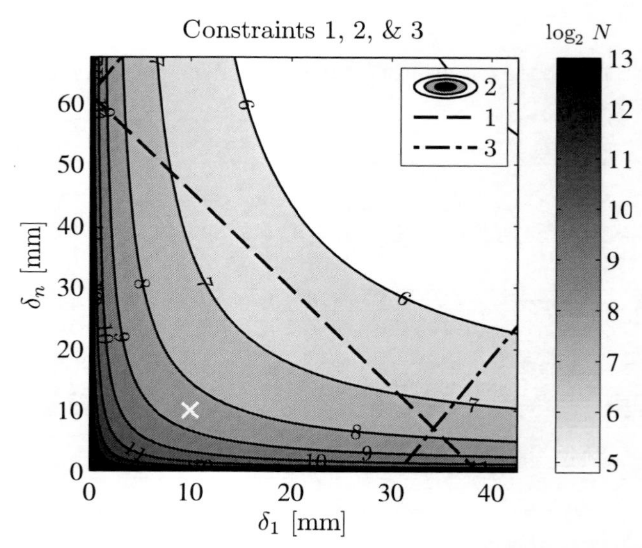

Figure 9.6 Graphical sampling analysis for the example point-source propagation. The region that satisfies constraint 1 is below the black dashed line, while the region above the black dash-dot line satisfies constraint 3. The white $\times$ marks the chosen values of δ_1 and δ_n.

9.5.3 Perform a vacuum simulation

With the grid parameters N, δ_1, and δ_n determined, the next step is to perform a vacuum simulation. This serves two important purposes. First, it verifies that the simulation is producing accurate results without regard to the turbulence. In this particular case, we are simulating a point source, so we can compare the vacuum simulation result against a known analytic solution. Listing 9.7 gives the MATLAB code that carries out a vacuum simulation for the example geometry. Lines 3–5 create copies of some variables from Listing 9.6. Then, lines 12–14 create the sinc-Gaussian model point source. Next, lines 19–25 setup and perform the propagation using a super-Gaussian absorbing boundary at each plane. Lastly, the computed field is collimated by removing the spherical-wave phase. This allows the phase difference to be studied, which is helpful for making some plots and absolutely necessary for certain analyses, like computing the coherence factor.

False-color, gray-scale images of the resulting irradiance and phase are shown in Fig. 9.7. Clearly, the irradiance in plot (a) is nearly uniform over the region of interest, and the phase in plot (b) is flat (after collimation). Plotting a slice of the phase with the theoretical expectation would reveal that the curvature is correct. The second purpose of performing a vacuum simulation is for comparison to the turbulent simulations. Often, we want to know how much the performance of an optical system is degraded by turbulence, so we need to know how the system performs in vacuum for comparison. This is necessary, for example, if we want to calculate the Strehl ratio.

Listing 9.7 MATLAB code for executing a vacuum simulation of the point source given the grid determined by sampling analysis.

```
1  % example_pt_source_vac_prop.m
2
3  delta1 = d1;        % source-plane grid spacing [m]
4  deltan = d2;        % observation-plane grid spacing [m]
5  n = nscr;             % number of planes
6
7  % coordinates
8  [x1 y1] = meshgrid((-N/2 : N/2-1) * delta1);
9  [theta1 r1] = cart2pol(x1, y1);
10
11 % point source
12 pt = exp(-i*k/(2*R) * r1.^2) / D1^2 ...
13     .* sinc(x1/D1) .* sinc(y1/D1) ...
14     .* exp(-(r1/(4*D1)).^2);
15 % partial prop planes
16 z = (1 : n-1) * Dz / (n-1);
17
18 % simulate vacuum propagation
19 sg = exp(-(x1/(0.47*N*d1)).^16) ...
20     .* exp(-(y1/(0.47*N*d1)).^16);
21 t = repmat(sg, [1 1 n]);
22 [xn yn Uvac] = ang_spec_multi_prop(pt, wvl, ...
23     delta1, deltan, z, t);
24 % collimate the beam
25 Uvac = Uvac .* exp(-i*pi/(wvl*R)*(xn.^2+yn.^2));
```

9.5.4 Perform the turbulent simulations

Finally, we can perform turbulent simulations with realizations of phase screens. Listing 9.8 gives the code for executing turbulent simulations for the example scenario. In the listing, we generate 11 phase screens (at the correct grid spacings, which may be different for each screen) to create one realization of a turbulent path and simulate the propagation. The process is repeated 40 times so that we have 40 realizations of optical fields propagated through independent and identically distributed atmospheres. A false-color, gray-scale image of one representative field is shown in Fig. 9.8 with the irradiance in plot (a) and phase in plot (b). Collecting many such realizations allows us to estimate ensemble statistics like the coherence factor, wave structure function, and log-amplitude variance.

If we wanted to simulate a dynamically evolving atmosphere, for each atmospheric realization we would need to move the phase screens in the transverse dimension as time evolves. This makes explicit use of the Taylor frozen-turbulence hypothesis.[15] The velocities of the screens needs to be determined from temporal quantities like the Greenwood frequency.[68] This would allow us to verify temporal properties of the simulation and then use the simulation with dynamic optical systems such as adaptive optics.

9.5.5 Verify the output

There are two simulation properties that are verified in this subsection. The first is the phase-screen structure function, and the second is the coherence factor of the observation-plane field. These verifications make use of independent and identically distributed realizations to check spatial correlations. If a dynamically evolving atmosphere is simulated, temporal properties like the temporal phase structure function should be checked as well.

First, the phase screens are verified. To do so, we can use the 40 random draws for any one partial-propagation plane. This is done by computing the two-

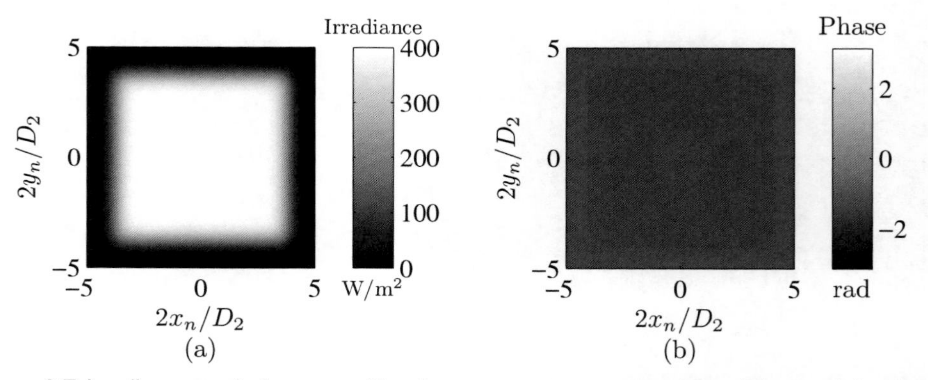

Figure 9.7 Irradiance and phase resulting from a vacuum propagation of the model point source. Note that line 25 of Listing 9.7 indicates that the field was collimated before plotting, which is visible in plot (b).

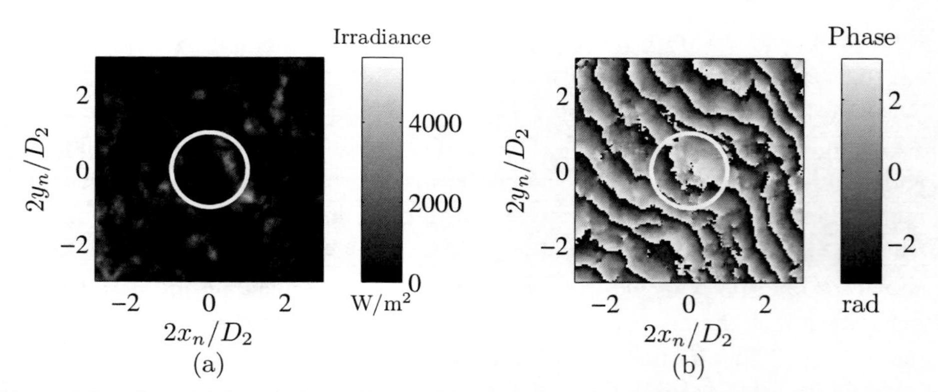

Figure 9.8 Irradiance (a) and phase (b) resulting from a turbulent propagation of the model point source. The white circle marks the edge of the observing telescope aperture. Note that the field was collimated before plotting, which is apparent in plot (b).

Listing 9.8 MATLAB code for executing a turbulent simulation of the point source given the grid determined by sampling analysis.

```
1  % example_pt_source_turb_prop.m
2
3  l0 = 0;       % inner scale [m]
4  L0 = inf;     % outer scale [m]
5
6  zt = [0 z]; % propagation plane locations
7  Delta_z = zt(2:n) - zt(1:n-1);     % propagation distances
8  % grid spacings
9  alpha = zt / zt(n);
10 delta = (1-alpha) * delta1 + alpha * deltan;
11
12 % initialize array for phase screens
13 phz = zeros(N, N, n);
14 nreals = 20;     % number of random realizations
15 % initialize arrays for propagated fields,
16 % aperture mask, and MCF
17 Uout = zeros(N);
18 mask = circ(xn/D2, yn/D2, 1);
19 MCF2 = zeros(N);
20 sg = repmat(sg, [1 1 n]);
21 for idxreal = 1 : nreals      % loop over realizations
22     idxreal
23     % loop over screens
24     for idxscr = 1 : 1 : n
25         [phz_lo phz_hi] ...
26             = ft_sh_phase_screen ...
27             (r0scrn(idxscr), N, delta(idxscr), L0, l0);
28         phz(:,:,idxscr) = phz_lo + phz_hi;
29     end
30     % simulate turbulent propagation
31     [xn yn Uout] = ang_spec_multi_prop(pt, wvl, ....
32         delta1, deltan, z, sg.*exp(i*phz));
33     % collimate the beam
34     Uout = Uout .* exp(-i*pi/(wvl*R)*(xn.^2+yn.^2));
35     % accumulate realizations of the MCF
36     MCF2 = MCF2 + corr2_ft(Uout, Uout, mask, deltan);
37 end
38 % modulus of the complex degree of coherence
39 MCDOC2 = abs(MCF2) / (MCF2(N/2+1,N/2+1));
```

dimensional structure function of each phase screen and then averaging each to obtain the mean structure function, as discussed in Sec. 3.3. Figure 9.9 shows an example comparison of the theoretical phase structure function from Eq. (9.44) to

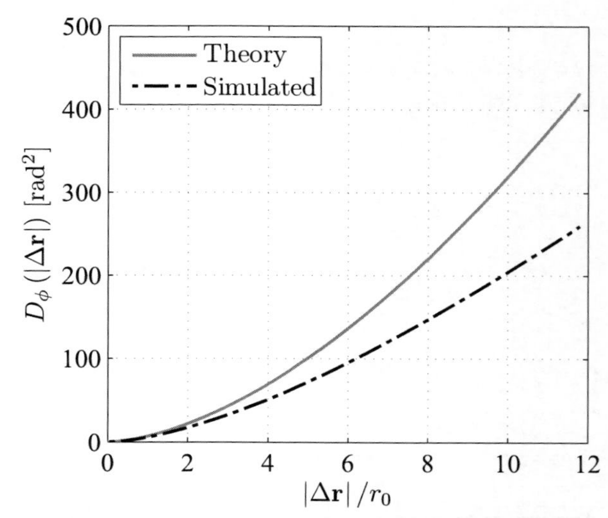

Figure 9.9 Verifying structure function of an ensemble of independent and identically distributed phase screens.

the average structure function computed from phase screen realizations. The comparison is close, indicating that the screens are adequately representing the phase accumulated along the propagation path.

To confirm that the turbulent simulation operates correctly, we have computed the coherence factor in the observation plane. Line 36 in Listing 9.8 accumulates the two-dimensional mutual coherence function using the `corr2_ft` function from Ch. 3, and line 39 normalizes to get the coherence factor. The result is plotted in Fig. 9.10 along with the theoretical expectation. The theoretical expectation combines Eqs. (9.32) and (9.44). We can see that there is a good match between theory and the simulation results. There is a slight departure, so if we need greater accuracy, we could go back to the setup and re-evaluate the choice of phase screen properties ro try an even more accurate screen generation method like the one developed by Johansson and Gavel.[88] One way to adjust the setup would be to examine Eq. (9.65) and adjust the values of z_i and Δz_i attempting to match turbulence moments of the continuous and layered models. The case of constant C_n^2 discussed here is a simple case for which uniformly spaced screens with uniform properties work fairly well. As an example of more extensive verification that could be performed, Martin and Flatté[43,44] tested their simulations by comparing the spatial irradiance PSD in the observation plane against weak turbulence theory and asymptotic theory.

9.6 Conclusion

The example given in this chapter has illustrated the steps that we must take to set up a simulation of optical propagation through turbulence and ensure accurate results. This is an important process, and many of these steps are often overlooked.

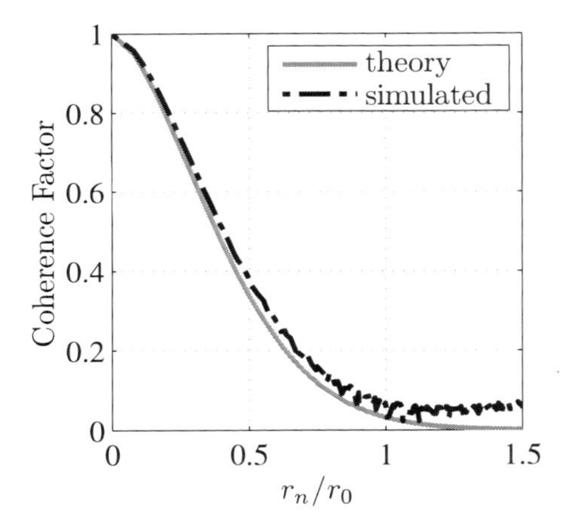

Figure 9.10 The coherence factor in the observation plane.

Because simulations can be much more complicated than the situation given here, often more effort is required to ensure accurate simulation results. Additional complexities often include two-way propagation, adaptive-optics systems, moving platforms, reflection from rough surfaces, multiple wavelengths, and much more.[36, 89] These additions need to be tested as thoroughly as the atmospheric propagation part of the simulation.

9.7 Problems

1. Show that if ϵ is position-dependent, Maxwell's equations combine similarly to the development in Sec. 1.2.1 to yield Eq. (9.20).

2. Show that for a propagation path with constant C_n^2, $r_{0,sw} = (3/8)^{-3/5}\, r_{0,pw}$.

3. Substitute Eq. (9.44) into Eq. (9.34) to show that Eq. (9.49) is the correct phase PSD for Kolmogorov turbulence.

4. Show that for a propagation path with constant C_n^2, $\sigma^2_{\chi,sw} = 0.404\sigma^2_{\chi,pw}$.

5. Show the sampling diagram for a point source with wavelength 1 μm propagating 2 km through an atmosphere with $r_0 = 2$ cm to a telescope with a 2-m-diameter aperture. Compare this to the vacuum case. How many more samples are needed? How many partial propagations are needed in each case?

6. Show the sampling diagram for a point source with wavelength 1 μm propagating 75 km through an atmosphere with $r_0 = 10$ cm to a telescope with a 1-m-diameter aperture. Compare this to the vacuum case. How many more samples are needed? How many partial propagations are needed in each case?

7. Consider propagating a point source with an optical wavelength of 1 μm a distance $\Delta z = 100$ km through an atmosphere with the Kolmogorov refractive-index PSD and $C_n^2 = 1 \times 10^{-17}$ m$^{-2/3}$ all along the path.

   (a) Analytically evaluate the integrals given in Eqs. (9.42), (9.43), (9.63), and (9.64) to compute the continuous-model r_0 and log-amplitude variance σ_χ^2 for both a plane wave and a point source, assuming that C_n^2 is constant along the propagation path.

   (b) Using three phase screens, write down the matrix-vector equations similar to Eq. (9.75) in an attempt to match the continuous and discrete point-source r_0, point-source log-amplitude variance, and plane-wave log-amplitude variance. Solve the system of equations for the three values of r_{0i}. With three parameters and three screens, there is a unique solution. Is it physically meaningful? Explain your answer.

   (c) Now, adapt the system of equations to accommodate seven phase screens and solve the system similarly to the method in Listing 9.5.

   (d) Given that the receiving aperture has a diameter of 2 m, perform the sampling analysis with consideration of the turbulence. Create a plot similar to Fig. 8.5

   (e) Generate the phase screens with 20 independent and identically distributed realizations using the Kolmogorov phase PSD. Compute the structure function for the last phase screen and plot it along with the appropriate theoretical expectation.

   (f) Simulate the propagation through the turbulent path and plot the coherence factor of the observation-plane field along with the theoretical expectation.

Appendix A
Function Definitions

Below are definitions of several functions used throughout the book. They are provided here so that the reader knows what conventions are being used for these functions.

The rectangle function (sometimes called the box function) is defined as

$$\operatorname{rect}\left(\frac{x}{a}\right) = \begin{cases} 1 & x < \frac{a}{2} \\ \frac{1}{2} & x = \frac{a}{2} \\ 0 & x > \frac{a}{2}. \end{cases} \tag{A.1}$$

The triangle function (sometimes called the hat or tent function) is defined as

$$\operatorname{tri}(ax) = \begin{cases} 1 - |ax| & |ax| < 1 \\ 0 & \text{otherwise.} \end{cases} \tag{A.2}$$

The sinc function is defined as

$$\operatorname{sinc}(ax) = \frac{\sin(a\pi x)}{a\pi x}. \tag{A.3}$$

The comb function (sometimes called the Shah function) is defined as

$$\operatorname{comb}(ax) = \sum_{n=-\infty}^{\infty} \delta(ax - n), \tag{A.4}$$

where $\delta(x)$ is the Dirac delta function.[90]

The circle function (sometimes called the cylinder function) is defined as

$$\operatorname{circ}\left(\frac{\sqrt{x^2+y^2}}{a}\right) = \begin{cases} 1 & \sqrt{x^2+y^2} < a \\ \frac{1}{2} & \sqrt{x^2+y^2} = a \\ 0 & \sqrt{x^2+y^2} > a. \end{cases} \tag{A.5}$$

The jinc function (sometimes called the besinc or sombrero function) is defined as

$$\operatorname{jinc}(ax) = 2\frac{J_1(a\pi x)}{a\pi x}, \tag{A.6}$$

where $J_n(x)$ is a Bessel function of the first kind of order n.[90]

Appendix B
MATLAB Code Listings

Below are MATLAB code listings for several functions used throughout the book. They are provided here so that the reader knows exactly how to generate samples of these signals.

Listing B.1 MATLAB code for evaluating the rect function.

```
1 function y = rect(x, D)
2 % function y = rect(x, D)
3     if nargin == 1, D = 1; end
4     x = abs(x);
5     y = double(x<D/2);
6     y(x == D/2) = 0.5;
```

Listing B.2 MATLAB code for evaluating the triangle function.

```
1 function y = tri(t)
2 % function y = tri(t)
3     t = abs(t);
4     y = zeros(size(t));
5     idx = find(t < 1.0);
6     y(idx) = 1.0 - t(idx);
```

Listing B.3 MATLAB code for evaluating the circ function.

```
1 function z = circ(x, y, D)
2 % function z = circ(x, y, D)
3     r = sqrt(x.^2+y.^2);
4     z = double(r<D/2);
5     z(r==D/2) = 0.5;
```

Listing B.4 MATLAB code for evaluating the jinc function.

```
1  function y = jinc(x)
2  % function y = jinc(x)
3      y = ones(size(x));
4      idx = x ~= 0;
5      y(idx) = 2.0*besselj(1, pi*x(idx)) ./ (pi*x(idx));
```

Listing B.5 MATLAB code for analytically evaluating the Fresnel diffraction pattern of a square aperture.

```
1  function U = fresnel_prop_square_ap(x2, y2, D1, wvl, Dz)
2  % function U = fresnel_prop_square_ap(x2, y2, D1, wvl, Dz)
3
4      N_F = (D1/2)^2 / (wvl * Dz); % Fresnel number
5      % substitutions
6      bigX = x2 / sqrt(wvl*Dz);
7      bigY = y2 / sqrt(wvl*Dz);
8      alpha1 = -sqrt(2) * (sqrt(N_F) + bigX);
9      alpha2 = sqrt(2) * (sqrt(N_F) - bigX);
10     beta1 = -sqrt(2) * (sqrt(N_F) + bigY);
11     beta2 = sqrt(2) * (sqrt(N_F) - bigY);
12     % Fresnel sine and cosine integrals
13     ca1 = mfun('FresnelC', alpha1);
14     sa1 = mfun('FresnelS', alpha1);
15     ca2 = mfun('FresnelC', alpha2);
16     sa2 = mfun('FresnelS', alpha2);
17     cb1 = mfun('FresnelC', beta1);
18     sb1 = mfun('FresnelS', beta1);
19     cb2 = mfun('FresnelC', beta2);
20     sb2 = mfun('FresnelS', beta2);
21     % observation-plane field
22     U = 1 /(2*i) *((ca2 - ca1) + i * (sa2 - sa1)) ...
23         .* ((cb2 - cb1) + i * (sb2 - sb1));
```

References

1. J. D. Jackson, *Classical Electrodynamics*, 3rd Ed., John Wiley & Sons, Inc., New York, NY (1998).
2. C. C. Davis, *Lasers and Electro-Optics: Fundamentals and Engineerng*, 3rd Ed., Cambridge University Press (1996).
3. J. Verdeyen, *Laser Engineering*, Prentice Hall (1994).
4. M. Born and E. Wolf, *Principles of Optics: Electromagnetic Theory of Propagation, Interference and Diffraction of Light*, 7th Ed., Cambridge University Press (1999).
5. J. W. Goodman, *Introduction to Fourier Optics*, 3rd Ed., Roberts & Co., Greenwood Village, CO (2005).
6. J. W. Goodman, *Statistical Optics*, John Wiley & Sons, Inc., New York, NY (1985).
7. The Mathworks, "MATLAB," 2007. Version 2007a.
8. E. O. Brigham, *Fast Fourier Transform and Its Applications*, Prentice Hall, Upper Saddle River, NJ (1998).
9. M. Frigo and S. G. Johnson, "The design and implementation of FFTW3," Proc. IEEE **93**(2), pp. 216–231 (2005). special issue on "Program Generation, Optimization, and Platform Adaptation".
10. W. H. Press, S. A. Teukolsky, W. T. Vetterling, and B. P. Flannery, *Numerical Recipes: The Art of Scientific Computing*, 3rd Ed., Cambridge University Press (2007).
11. Visual Numerics, Inc., "IMSL Numerical Libraries." computer software.
12. B. Sklar, *Digital Communications: Fundamentals and Applications*, 2nd Ed., Prentice Hall, Upper Saddle River, NJ (2001).
13. J. F. James, *A Student's Guide to Fourier Transforms: with Applications in Physics and Engineering*, 2nd Ed., Cambridge University Press, Cambridge, UK (2002).
14. F. G. Stremler, *Introduction to Communication Systems*, 3rd Ed., Prentice Hall (1990).
15. L. C. Andrews and R. L. Phillips, *Laser Beam Propagation Through Random Media*, 2nd Ed., SPIE Press, Bellingham, WA (2005).
16. Optical Research Associates, "CODE V." computer software.
17. Lambda Research Corporation, "OSLO." computer software.
18. ZEMAX Development Corporation, "ZEMAX." computer software.

19. V. N. Mahajan, *Optical Imaging and Aberrations Part II: Wave Diffraction Optics*, SPIE Press, Bellingham, WA (1998).

20. C. Zhao and J. H. Burge, "Orthonormal vector polynomials in a unit circle, Part I: basis set derived from gradients of Zernike polynomials," *Opt. Express* **15**(26), pp. 18014–18024 (2007).

21. C. Zhao and J. H. Burge, "Orthonormal vector polynomials in a unit circle, Part II : completing the basis set," *Opt. Express* **16**(9), pp. 6586–6591 (2008).

22. R. Noll, "Zernike polynomials and atmospheric turbulence," *J. Opt. Soc. Am.* **66**, pp. 207–211 (1976).

23. M. C. Roggemann and B. M. Welsh, *Imaging Through Turbulence*, CRC Press, Inc., New York, NY (1996).

24. R. Navarro, J. Arines, and R. Rivera, "Direct and inverse discrete Zernike transform," *Opt. Express* **17**(26), pp. 24269–24281 (2009).

25. E. Anderson, Z. Bai, C. Bischof, J. Demmel, J. Dongarra, J. D. Croz, A. Greenbaum, S. Hammarling, A. McKenney, and D. Sorenson, "LAPACK: A portable linear algebra library for high-performance computers," Tech. Rep. CS-90-105, University of Tennessee, Knoxville, TN (1990).

26. J. Dongarra, "Basic linear algebra subprograms technical forum standard," *International Journal of High Performance Applications and Supercomputing* **16**(1), pp. 1–111 (2002).

27. J. Dongarra, "Basic linear algebra subprograms technical forum standard," *International Journal of High Performance Applications and Supercomputing* **16**(2), pp. 115–199 (2002).

28. N. Delen and B. Hooker, "Free-space beam propagation between arbitrarily oriented planes based on full diffraction theory: a fast Fourier transform approach," *J. Opt. Soc. Am. A* **15**(4), pp. 857–867 (1998).

29. N. Delen and B. Hooker, "Verification and comparison of a fast Fourier transform-based full diffraction method for tilted and offset planes," *Appl. Opt.* **40**(21), pp. 3525–3531 (2001).

30. G. A. Tyler and D. L. Fried, "A wave optics propagation algorithm," Tech. Rep. TR-451, the Optical Sciences Company (1982).

31. P. H. Roberts, "A wave optics propagation code," Tech. Rep. TR-760, the Optical Sciences Company (1986).

32. W. A. Coles, J. P. Filice, R. G. Frehlich, and M. Yadlowsky, "Simulation of wave propagation in three-dimensional random media," *Appl. Opt.* **34**(12), pp. 2089–2101 (1995).

33. J. A. Rubio, A. Belmonte, and A. Comerón, "Numerical simulation of long-path spherical wave propagation in three-dimensional random media," *J. Opt. Soc. Am. A* **38**(9), pp. 1462–1469 (1999).

34. X. Deng, B. Bihari, J. Gan, F. Zhao, and R. T. Chen, "Fast algorithm for chirp transforms with zooming-in ability and its applications," *J. Opt. Soc. Am. A* **17**(4), pp. 762–771 (2000).

35. S. Coy, "Choosing mesh spacings and mesh dimensions for wave optics simulation," Proc. SPIE **5894**, (2005).

36. C. Rydberg and J. Bengtsson, "Efficient numerical representation of the optical field for the propagation of partially coherent radiation with a specified spatial and temporal coherence function," *J. Opt. Soc. Am. A* **23**(7), pp. 1616–1625 (2006).

37. D. G. Voelz and M. C. Roggemann, "Digital simulation of scalar optical diffraction: revisiting chirp function sampling criteria and consequences," *Appl. Opt.* **48**(32), pp. 6132–6142 (2009).

38. M. Nazarathy and J. Shamir, "Fourier optics described by operator algebra," *J. Opt. Soc. Am. A* **70**(2), pp. 150–159 (1980).

39. M. Nazarathy and J. Shamir, "First-order optics-a canonical operator representation: lossless systems," *J. Opt. Soc. Am. A* **72**(3), pp. 356–364 (1982).

40. J. M. Jarem and P. P. Banerjee, *Computational Methods for Electromagnetic and Optical Systems*, Marcel Dekker, Inc., New York, NY (2000).

41. R. A. Johnston and R. G. Lane, "Modeling scintillation from an aperiodic Kolmogorov phase screen," *Appl. Opt.* **39**(26), pp. 4761–4769 (2000).

42. J. D. Mansell, R. Praus, and S. Coy, "Determining wave-optics mesh parameters for complex optical systems," Proc. SPIE **6675** (2007).

43. J. M. Martin and S. M. Flatté, "Intensity images and statistics from numerical simulation of wave propagation in 3-D random media," *Appl. Opt.* **27**(11), pp. 2111–2126 (1988).

44. J. M. Martin and S. M. Flatté, "Simulation of point-source scintillation through three-dimensional random media," *J. Opt. Soc. Am. A* **7**(5), pp. 838–847 (1990).

45. F. L. Pedrotti, L. M. Pedrotti, and L. S. Pedrotti, *Introduction to Optics*, 3rd Ed., Benjamin Cummings (2006).

46. C. Palma and V. Bagini, "Extension of the Fresnel transform to ABCD systems," *J. Opt. Soc. Am. A* **14**(8), pp. 1774–1779 (1997).

47. A. J. Lambert and D. Fraser, "Linear systems approach to simulating optical diffraction," *Appl. Opt.* **37**(34), pp. 7933–7939 (1998).

48. J. D. Mansell, L. Xu, A. S. amd Robert Praus, and S. Coy, "Algorithm for implementing an ABCD ray matrix wave-optics propagator," Proc. SPIE **6675** (2007).

49. H. M. Ozaktas and D. Mendlovic, "Fractional Fourier optics," *J. Opt. Soc. Am. A* **12**(4), pp. 743–750 (1995).

50. J. García, D. Mas, and R. G. Dorsch, "Fractional Fourier transform calculation through the fast-Fourier-transform algorithm," *Appl. Opt.* **35**(35), pp. 7013–7018 (1996).

51. F. J. Marinho and L. M. Bernardo, "Numerical calculation of fractional Fourier transforms with a single fast-Fourier-transform algorithm," *J. Opt. Soc. Am. A* **15**(8), pp. 2111–2116 (1998).

52. D. Mas, J. García, C. Ferreira, L. M. Bernardo, and F. J. Marinho, "Fast algorithms for free-space diffraction patterns calculation," *Opt. Commun.* **164**(4), pp. 233–245 (1999).

53. S. M. Flatté, G.-Y. Wang, and J. Martin, "Irradiance variance of optical waves through atmospheric turbulence by numerical simulation and comparison with experiment," *J. Opt. Soc. Am. A* **10**(11), pp. 2363–2370 (1993).

54. S. Coy, "How to choose mesh spacings for wave-optics simulations," tech. rep., MZA Associates (2003).

55. L. Onural, "Some mathematical properties of the uniformly sampled quadratic phase function and associated issues in digital Fresnel diffraction simulations," *Opt. Eng.* **43**(11), pp. 2557–2563 (2004).

56. R. Frehlich, "Simulation of laser propagation in a turbulent atmosphere," *Appl. Opt.* **39**(3), pp. 393–397 (2000).

57. T.-C. Poon and P. P. Banerjee, *Contemporary Optical Image Processing With Matlab*, Elsevier Science, Ltd., Oxford, UK (2001).

58. T.-C. Poon and T. Kim, *Engineering Optics with* MATLAB, World Scientific Publishing Co. (2006).

59. V. P. Lukin and B. V. Fortes, *Adaptive Beaming and Imaging in the Turbulent Atmosphere*, SPIE Press, Bellingham, WA (2002).

60. S. V. Mantravadi, T. A. Rhoadarmer, and R. S. Glas, "Simple laboratory system for generating well-controlled atmospheric-like turbulence," Proc. SPIE **5553** (2004).

61. T. A. Rhoadarmer and R. P. Angel, "Low-cost, broadband static phase plate for generating atmosphericlike turbulence," *Appl. Opt.* **40**, pp. 2946–2955 (2001).

62. A. N. Kolmogorov, "The local structure of turbulence in an incompressible viscous fluid for very large Reynolds numbers," *C. R. (Doki) Acad. Sci. U.S.S.R.* **30**, pp. 301–305 (1941).

63. A. M. Obukhov, "Structure of the temperature field in turbulent flow," *Izv. Acad. Nauk. SSSR, Ser. Georgr. I Geofiz.* **13**, pp. 58–69 (1949).

64. S. Corrsin, "On the spectrum of isotropic temperature fluctuations in isotropic turbulence," *J. Appl. Phys.* **22**, pp. 469–473 (1951).

65. A. Ishimaru, *Wave Propagation and Scattering in Random Media*, Wiley-IEEE Press, New York, NY (1999).

66. A. D. Wheelon, *Electromagnetic Scintillation: Volume 2, Weak Scattering*, Cambridge University Press (2003).

67. S. F. Clifford, *Laser Beam Propagation in the Atmosphere*, ch. The Classical Theory of Wave Propagation in the Atmosphere. Springer-Verlag (1978).

68. R. J. Sasiela, *Electromagnetic Wave Propagation in Turbulence: Evaluation and Application of Mellin Transforms*, 2nd Ed., SPIE Press, Bellingham, WA (2007).

69. D. L. Fried, "Statistics of a geometric representation of wavefront distortion," *J. Opt. Soc. Am.* **55**(11), pp. 1427–1431 (1965).

70. L. C. Andrews, S. Vester, and C. E. Richardson, "Analytic expressions for the wave structure function based on a bump spectral model for refractive index fluctuations," *J. Mod. Opt.* **40**, pp. 931–938 (1993).

71. M. C. Roggemann, B. M. Welsh, D. Montera, and T. A. Rhoadarmer, "Method for simulating atmospheric turbulence phase effects for multiple time slices and anisoplanatic conditions," *Appl. Opt.* **34**(20), pp. 4037–4051 (1995).

72. C. M. Harding, R. A. Johnston, and R. G. Lane, "Fast simulation of a Kolmogorov phase screen," *Appl. Opt.* **38**(11), pp. 2161–2170 (1999).

73. F. Assémat, R. W. Wilson, and E. Gendron, "Method for simulating infinitely long and non stationary phase screens with optimized memory storage," *Opt. Express* **14**(3), pp. 988–999 (2006).

74. A. Beghi, A. Cenedese, and A. Masiero, "Stochastic realization approach to the efficient simulation of phase screens," *J. Opt. Soc. Am. A* **25**(2), pp. 515–525 (2008).

75. V. Sriram and D. Kearney, "An ultra fast Kolmogorov phase screen generator suitable for parallel implementation," *Opt. Express* **15**(21), pp. 13709–13714 (2007).

76. G. Cochran, "Phase screen generation," Tech. Rep. TR-663, the Optical Sciences Company (1982).

77. B. J. Herman and L. A. Strugala, "Method for inclusion of low-frequency contributions in numerical representation of atmospheric turbulence," Proc. SPIE **1221**, pp. 183–192 (1990).

78. R. G. Lane, A. Glindemann, , and J. C. Dainty, "Simulation of a Kolmogorov phase screen," *Waves in Random Media* **2**, pp. 209–224 (1992).

79. H. Jakobssen, "Simulations of time series of atmospherically distorted wave fronts," *Appl. Opt.* **35**, pp. 1561–1565 (1996).

80. B. M. Welsh, "A Fourier series based atmospheric phase screen generator for simulating anisoplanatic geometries and temporal evolution," Proc. SPIE **3125**, pp. 327–338 (1997).

81. G. Sedmak, "Performance analysis of and compensation for aspect-ratio effects of fast-Fourier-transform-based simulations of large atmospheric wave fronts," *Appl. Opt.* **37**, pp. 4605–4613 (1998).

82. R. J. Eckert and M. E. Goda, "Polar phase screens: a comparative analysis with other methods of random phase screen generation," Proc. SPIE **6303** (2006).

83. D. Kouznetsov, V. V. Voitsekhovich, and R. Ortega-Martinez, "Simulations of turbulence-induced phase and log-amplitude distortions," *Appl. Opt.* **36**, pp. 464–469 (1997).

84. F. Dios, J. Recolons, A. Rodríguez, and O. Batet, "Temporal analysis of laser beam propagation in the atmosphere using computer-generated long phase screens," *Opt. Express* **16**(3), pp. 2206–2220 (2008).

85. D. L. Fried and T. Clark, "Extruding Kolmogorov-type phase screen ribbons," *J. Opt. Soc. Am. A* **25**(2), pp. 463–468 (2008).

86. B. L. McGlamery, "Restoration of turbulence-degraded images," *J. Opt. Soc. Am.* **57**(3), pp. 293–297 (1967).

87. N. A. Roddier, "Atmospheric wavefront simulation using Zernike polynomials," *Opt. Eng.* **29**, pp. 1174–1180 (1990).

88. E. M. Johansson and D. T. Gavel, "Simulation of stellar speckle imaging,"Proc. SPIE **2200**, pp. 372–383 (1994).

89. G. J. Gbur, "Simulating fields of arbitrary spatial and temporal coherence," *Opt. Express* **14**(17), pp. 7567–7578 (2006).

90. H. J. Weber and G. B. Arfken, *Mathematical Methods for Physicists*, 6th Ed., Academic Press (2005).

Index

Jason D. Schmidt is a Major in the U.S. Air Force and an assistant professor of electro-optics at the Air Force Institute of Technology in the Department of Electrical and Computer Engineering. Previously, he was a research physicist at the U.S. Air Force Research Laboratory's Starfire Optical Range. He received the doctoral degree in Electro-Optics from the University of Dayton. Dr. Schmidt has been an active researcher in optical wave propagation through atmospheric turbulence for ten years. He received the Young Investigator Award in 2008 from the Air Force Office of Scientific Research. Besides optical wave propagation, Dr. Schmidt's research interests include free-space optical communications and adaptive optics.